THE AZAR-HAGEN GRAMMAR SERIES

TEST BANK FOR FUNDAMENTALS OF English Grammar

FIFTH EDITION

Kelly Roberts Weibel

Fundamentals of English Grammar, Fifth Edition
Test Bank

Copyright © 2020, 2011, 2003 by Pearson Education, Inc.
All rights reserved.

No part of this publication may be reproduced,
stored in a retrieval system, or transmitted in any
form or by any means, electronic, mechanical,
photocopying, recording, or otherwise, without the
prior permission of the publisher.

Azar Associates: Sue Van Etten, Manager

Pearson Education, 221 River Street, Hoboken, NJ 07030

Staff credits: The people who made up the *Fundamentals of English Grammar, Fifth Edition, Test Bank* team, representing content development, project management, design, and publishing are Pietro Alongi, Sheila Ameri, Warren Fischbach, Sarah Henrich, Niki Lee, Amy McCormick, Robert Ruvo, and Paula Van Ells.

Contributing Editor: Lise Minovitz
Text composition: Integra

Printed in the United States of America
ISBN 10: 0-13-563580-2
ISBN 13: 978-0-13-563580-3
1 2019

CONTENTS

INTRODUCTION viii

CHAPTER 1 PRESENT TIME 1
- Quiz 1   Simple Present and Present Progressive 1
- Quiz 2   Simple Present Tense 1
- Quiz 3   Present Progressive Tense 2
- Quiz 4   Questions and Negatives in the Present 2
- Quiz 5   Singular or Plural 3
- Quiz 6   Frequency Adverbs 4
- Quiz 7   Simple Present vs. Present Progressive 5
- Quiz 8   Verbs Not Usually Used in the Progressive 5
- Quiz 9   Simple Present vs. Present Progressive 6
- Quiz 10 Present Verbs: Short Answers to *Yes / No* Questions 6
- Quiz 11 Chapter Review 7
- Quiz 12 Chapter Review 8
- Quiz 13 Chapter Review 9
- Chapter 1 - Test 1 10
- Chapter 1 - Test 2 12

CHAPTER 2 PAST TIME 14
- Quiz 1   Simple Past Tense 14
- Quiz 2   Simple Past: Negatives 15
- Quiz 3   Simple Past: Questions 16
- Quiz 4   Simple Past: Negatives and Questions 16
- Quiz 5   Simple Past: Questions and Answers 17
- Quiz 6   Spelling of *-ing* and *-ed* Forms 18
- Quiz 7   Simple Past 19
- Quiz 8   Simple Past 20
- Quiz 9   Simple Past 20
- Quiz 10 The Past Progressive 21
- Quiz 11 Understanding Past Time 21
- Quiz 12 Simple Past vs. Past Progressive 22
- Quiz 13 Simple Past and Past Progressive 23
- Quiz 14 Simple Past and Past Progressive 24
- Quiz 15 Expressing Past Time: Understanding Time Clauses 24
- Quiz 16 Expressing Past Time: Using Time Clauses 26
- Quiz 17 Expressing Past Habit: *Used To* 27
- Quiz 18 Chapter Review 27
- Quiz 19 Chapters 1 and 2 Review 28
- Chapter 2 - Test 1 29
- Chapter 2 - Test 2 31

CHAPTER 3 FUTURE TIME 33
- Quiz 1   Understanding Present, Past, and Future 33
- Quiz 2   *Will* and *Be Going To* 33
- Quiz 3   Questions with *Will* and *Be Going To* 34
- Quiz 4   *Be Going To* 35
- Quiz 5   *Will* 36
- Quiz 6   Contractions with *Will* and *Be Going To* 36
- Quiz 7   *Be Going To* vs. *Will* 37
- Quiz 8   *Be Going To* vs. *Will* 38
- Quiz 9   Certainty About the Future 39
- Quiz 10 Certainty About the Future 39
- Quiz 11 Expressing the Future in Time Clauses and *If*-Clauses 40
- Quiz 12 Using *Be Going To* and the Present Progressive to Express Future Time 41

Quiz 13	Using the Simple Present to Express Future Time	42
Quiz 14	Expressing Future Time	43
Quiz 15	Chapter Review	43
Quiz 16	Chapter Review	44
Quiz 17	Chapters 1 → 3 Review	45

Chapter 3 - Test 1 46
Chapter 3 - Test 2 48

CHAPTER 4 PRESENT PERFECT AND PAST PERFECT 50

Quiz 1	The Past Participle	50
Quiz 2	Present Perfect with Unspecified Time: *Ever* and *Never*	51
Quiz 3	Present Perfect with Unspecified Time: *Already* and *Yet*	52
Quiz 4	Present Perfect with *Since* and *For*	53
Quiz 5	Present Perfect with *Since* and *For*	54
Quiz 6	*Since* vs. *For*	54
Quiz 7	*Since* vs. *For*	55
Quiz 8	Time Clauses with *Since*	55
Quiz 9	Simple Past vs. Present Perfect	56
Quiz 10	Simple Past vs. Present Perfect	57
Quiz 11	Present Perfect Progressive	57
Quiz 12	Present Progressive vs. Present Perfect Progressive	59
Quiz 13	Present Progressive vs. Present Perfect Progressive	59
Quiz 14	Present Perfect Progressive vs. Present Perfect	60
Quiz 15	Past Perfect	61
Quiz 16	Past Perfect	62
Quiz 17	Chapter Review	63
Quiz 18	Chapter 1 → 4 Review	64

Chapter 4 - Test 1 66
Chapter 4 - Test 2 68

CHAPTER 5 ASKING QUESTIONS 70

Quiz 1	*Yes / No* Questions and Short Answers	70
Quiz 2	*Yes / No* Questions and Short Answers	71
Quiz 3	*Where, Why, When,* and *What Time*	72
Quiz 4	*Where, Why, When, What Time, How Come,* and *What ... For*	73
Quiz 5	Questions with *Who* and *What*	74
Quiz 6	Questions with *Who* and *What*	75
Quiz 7	Using *What* and a Form of *Do*	76
Quiz 8	*Which* vs. *What*	77
Quiz 9	Questions with *How*	78
Quiz 10	Questions with *How Often, How Many, How Far,* and *How Long*	79
Quiz 11	Review of *How*	80
Quiz 12	Information Questions Review	81
Quiz 13	Tag Questions	82
Quiz 14	Tag Questions	82
Quiz 15	Chapter Review	83

Chapter 5 - Test 1 84
Chapter 5 - Test 2 86

CHAPTER 6 NOUNS AND PRONOUNS 88

Quiz 1	Plural Forms of Nouns	88
Quiz 2	Pronunciation of Final *-s / -es*	88
Quiz 3	Subjects, Verbs, and Objects	89
Quiz 4	Subjects, Verbs, and Objects	90
Quiz 5	Objects of Prepositions	90
Quiz 6	Prepositions of Time	91
Quiz 7	Subject and Verb Agreement	91
Quiz 8	Using Adjectives	92

	Quiz 9	Using Adjectives	92
	Quiz 10	Using Nouns as Adjectives	93
	Quiz 11	Personal Pronouns: Subjects and Objects	93
	Quiz 12	Possessive Nouns	94
	Quiz 13	*Who's* vs. *Whose*	95
	Quiz 14	Possessive Pronouns and Adjectives	96
	Quiz 15	Reflexive Pronouns	96
	Quiz 16	*Another* vs. *The Other*	97
	Quiz 17	*Other(s)* vs. *The Other(s)*	97
	Quiz 18	Forms of *Other*	98
	Quiz 19	Chapter Review	99
	Chapter 6 - Test 1		100
	Chapter 6 - Test 2		102

CHAPTER 7   MODAL AUXILIARIES, THE IMPERATIVE, MAKING SUGGESTIONS, STATING PREFERENCES ... 104

	Quiz 1	The Form of Modal Auxiliaries	104
	Quiz 2	Expressing Ability	104
	Quiz 3	Possibility vs. Permission	105
	Quiz 4	Expressing Possibility	106
	Quiz 5	Meanings of *Could*	107
	Quiz 6	Ability, Possibility, and Permission	108
	Quiz 7	Polite Questions	109
	Quiz 8	*Should* and *Ought To*	110
	Quiz 9	*Should / Ought To* vs. *Had Better*	111
	Quiz 10	Necessity: *Have To, Have Got To,* and *Must*	112
	Quiz 11	*Do Not Have To* and *Must Not*	112
	Quiz 12	Review of *Must, Have To,* and *Had To*	113
	Quiz 13	Logical Conclusion vs. Necessity	114
	Quiz 14	Making Logical Conclusions: *Must*	114
	Quiz 15	Tag Questions with Modals	115
	Quiz 16	Imperative Sentences: Giving Instructions	116
	Quiz 17	Making Suggestions: *Let's* and *Why Don't*	117
	Quiz 18	*Prefer, Like ... Better,* and *Would Rather*	117
	Quiz 19	Chapter Review	118
	Chapter 7 - Test 1		120
	Chapter 7 - Test 2		123

CHAPTER 8   CONNECTING IDEAS: PUNCTUATION AND MEANING ... 125

	Quiz 1	Punctuating with Commas and Periods	125
	Quiz 2	Connecting Ideas: *And, But,* and *Or*	125
	Quiz 3	*So* vs. *But*	126
	Quiz 4	Using Auxiliary Verbs After *But*	126
	Quiz 5	Auxiliary Verbs after *And* and *But*	127
	Quiz 6	Using *And* + *Too, So, Either,* and *Neither*	128
	Quiz 7	Connecting Ideas with *Because*	129
	Quiz 8	Connecting Ideas: *So* and *Because*	129
	Quiz 9	Connecting Ideas: *Even Though / Although* and *Because*	130
	Quiz 10	Punctuating Adverb Clauses	132
	Quiz 11	Chapter Review	132
	Chapter 8 - Test 1		133
	Chapter 8 - Test 2		135

CHAPTER 9   COMPARISONS ... 137

	Quiz 1	Comparatives	137
	Quiz 2	Comparatives	137
	Quiz 3	Superlatives	138
	Quiz 4	Completing Comparatives	139

	Quiz 5	Completing Superlatives	140
	Quiz 6	Comparatives vs. Superlatives	141
	Quiz 7	Comparatives with Adverbs	141
	Quiz 8	*Farther* and *Further*	142
	Quiz 9	Repeating a Comparative	142
	Quiz 10	Using Double Comparatives	143
	Quiz 11	Modifying Comparatives	144
	Quiz 12	Negative Comparisons	145
	Quiz 13	Comparisons with *As ... As*	146
	Quiz 14	Comparisons with *As ... As*	147
	Quiz 15	*Less ... Than* and *Not As ... As*	148
	Quiz 16	Comparatives with Nouns, Adjectives, and Adverbs	148
	Quiz 17	*The Same, Similar, Different, Like,* and *Alike*	149
	Quiz 18	*The Same, Similar, Different, Like,* and *Alike*	150
	Quiz 19	Chapter Review	151
	Chapter 9 - Test 1		152
	Chapter 9 - Test 2		155
CHAPTER 10	**THE PASSIVE**		**158**
	Quiz 1	Active or Passive	158
	Quiz 2	Forming the Passive	158
	Quiz 3	Forming the Passive	159
	Quiz 4	Understanding the Passive	160
	Quiz 5	Active vs. Passive	161
	Quiz 6	Progressive Forms of the Passive	162
	Quiz 7	Transitive and Intransitive Verbs	163
	Quiz 8	Transitive vs. Intransitive	163
	Quiz 9	Using the *By*-Phrase	164
	Quiz 10	Passive Modals	165
	Quiz 11	Review: Active vs. Passive	166
	Quiz 12	Past Participles as Adjectives	167
	Quiz 13	Past Participles as Adjectives	167
	Quiz 14	Participial Adjectives: *-ed* vs. *-ing*	168
	Quiz 15	Participial Adjectives: *-ed* vs. *-ing*	168
	Quiz 16	*Get* + Adjective; *Get* + Past Participle	169
	Quiz 17	*Be Used To / Be Accustomed To*	170
	Quiz 18	*Used To* vs. *Be Used To*	170
	Quiz 19	*Be Supposed To*	171
	Quiz 20	Chapter Review	172
	Chapter 10 - Test 1		174
	Chapter 10 - Test 2		177
CHAPTER 11	**COUNT / NONCOUNT NOUNS AND ARTICLES**		**180**
	Quiz 1	*A* vs. *An*	180
	Quiz 2	*A / An* vs. *Some*	180
	Quiz 3	Count and Noncount Nouns	181
	Quiz 4	*Much / Many*	181
	Quiz 5	*A Few / A Little*	182
	Quiz 6	*A Lot Of, Some, Several, Many / Much,* and *A Few / A Little*	183
	Quiz 7	Nouns That Can Be Count or Noncount	183
	Quiz 8	Units of Measure with Noncount Nouns	184
	Quiz 9	Understanding Articles with Count and Noncount Nouns	184
	Quiz 10	Articles: Specific vs. Non-specific	185
	Quiz 11	*The* vs. *A / An*	185
	Quiz 12	More About Articles: Using *The*	186
	Quiz 13	*The* for Second Mention	187
	Quiz 14	Using *The* or *Ø*	187
	Quiz 15	Review: *A, An, Ø,* and *The*	188

	Quiz 16	*The* or Ø with People and Places	189
	Quiz 17	Capitalization	189
	Quiz 18	Chapter Review	190
	Chapter 11 - Test 1		191
	Chapter 11 - Test 2		193

CHAPTER 12 ADJECTIVE CLAUSES ... 195

	Quiz 1	Using *Who* and *That* to Describe People	195
	Quiz 2	*Who* vs. *Whom in Adjective Clauses*	196
	Quiz 3	Using *Who, That, Ø,* and *Whom* to Describe People	196
	Quiz 4	Using *That* and *Which* to Describe Things	197
	Quiz 5	Review of Adjective Clauses	198
	Quiz 6	Singular and Plural Verbs in Adjective Clauses	199
	Quiz 7	Prepositions in Adjective Clauses	199
	Quiz 8	Prepositions in Adjective Clauses	200
	Quiz 9	Using *Whose* in Adjective Clauses	201
	Quiz 10	Chapter Review	202
	Chapter 12 - Test 1		203
	Chapter 12 - Test 2		205

CHAPTER 13 GERUNDS AND INFINITIVES ... 207

	Quiz 1	Verb + Gerund	207
	Quiz 2	*Go* + *-ing*	208
	Quiz 3	Verb + Gerund or Infinitive	208
	Quiz 4	Verb + Gerund or Infinitive	209
	Quiz 5	Preposition + Gerund	210
	Quiz 6	*By* vs. *With*	211
	Quiz 7	*By* + Gerund	211
	Quiz 8	Gerunds as Subjects and *It* + Infinitive	212
	Quiz 9	Expressing Purpose: *In Order To*	213
	Quiz 10	Expressing Purpose: *To* vs. *For*	214
	Quiz 11	Using Infinitives with *Too* and *Enough*	215
	Quiz 12	Chapter Review	216
	Chapter 13 - Test 1		218
	Chapter 13 - Test 2		220

CHAPTER 14 NOUN CLAUSES ... 222

	Quiz 1	Identifying Noun Clauses	222
	Quiz 2	Noun Clauses with Question Words	222
	Quiz 3	Noun Clauses and Information Questions	223
	Quiz 4	Noun Clauses with *If* and *Whether*	224
	Quiz 5	Noun Clauses with *That*	225
	Quiz 6	Substituting *So* or *Not* for a *That*-Clause	225
	Quiz 7	Quoted Speech	226
	Quiz 8	Quoted Speech	227
	Quiz 9	Quoted Speech vs. Reported Speech	227
	Quiz 10	Verb Forms in Reported Speech	228
	Quiz 11	Reported Speech	229
	Quiz 12	*Say, Tell,* and *Ask*	230
	Quiz 13	Chapter Review	230
	Chapter 14 - Test 1		231
	Chapter 14 - Test 2		233

MIDTERM & FINAL EXAMS ... 235

Midterm Exam 1: Chapters 1-7	235
Midterm Exam 2: Chapters 1-7	241
Final Exam 1: Chapters 1-14	244
Final Exam 2: Chapters 1-14	249

ANSWER KEY ... 252

This test bank accompanies *Fundamentals of English Grammar, Fifth Edition*. Instructors can choose from over 200 Quizzes and thirty-two tests to use for assessment. Teachers familiar with the third edition will find a great deal of material has been updated for this fourth edition.

QUIZZES

Each chapter contains a series of quizzes keyed to individual charts in the student book, followed by two chapter tests. The quizzes are intended as quick checks of student understanding for both teacher and student. Mastery of a quiz is a strong indicator that students are ready to progress to the next section.

CHAPTER TESTS

The tests at the end of each chapter are comprehensive, covering as many points from the chapter as possible. The formats of the questions in the chapter tests follow those used in the previous quizzes. The two chapter tests are identical in format so that one may be used as a practice test if desired.

EXAMS

Two midterm exams covering chapters one through seven and two comprehensive final exams are included in this test bank. They can be used in conjunction with the other quizzes and tests or used separately.

FORMAT

Because students bring a variety of learning styles to the classroom, there is a wide selection of test formats, including sentence completion, sentence connection, multiple choice, and error analysis, as well as more open completion. To maximize the use of the answer key, open-ended writing practice has been kept to a minimum. Teachers wishing to incorporate more writing into the tests are encouraged to add their own material at the end of the chapter tests.

ANSWER KEY

An answer key for all quizzes, tests, and exams can be found in the back of the text.

DUPLICATION

The material has been formatted so teachers can easily make copies for their students. Permission is granted to duplicate as many copies as needed for classroom use only.

Acknowledgments

I am forever grateful to my husband, Pat, for his continued support for the duration of this project. I have enjoyed working on this edition with my insightful and good-humored editor, Lise Minovitz. As always, I have appreciated the opportunity to collaborate once again with Stacy Hagen and Ruth Voetmann, whom I am lucky to call friends as well as colleagues.

Kelly Roberts Weibel

CHAPTER 1 Present Time

QUIZ 1 Simple Present and Present Progressive (Charts 1-1 and 1-2)

Directions: Complete the chart with the correct forms of the verb **walk**. The first one is done for you.

Simple Present Tense	Present Progressive Tense
I _____walk_____ .	I _____am walking_____ .
You _____ .	You _____ .
Laura _____ .	Henri _____ .
We _____ .	My husband and I _____ .
You and your classmates _____ .	You and your sister _____ .
They _____ .	Suzy and Charles _____ .

QUIZ 2 Simple Present Tense (Charts 1-1 and 1-2)

Directions: Complete the sentences. Use the simple present form of the verbs in parentheses.

SITUATION: *A Weekday Morning*

Example: The coffee shop on Sixth Avenue (*serve*) _____serves_____ the best coffee in town.

1. Ben (*eat*) _____ breakfast at 6:00 every morning.
2. Suzie (*eat, not*) _____ any breakfast.
3. Her neighbor, Mrs. Jones, (*enjoy*) _____ her breakfast on her patio.
4. Many children in our neighborhood (*walk*) _____ to school every day.
5. Some parents (*take*) _____ their younger children to school.
6. A few teenagers (*drive*) _____ their own cars to school.
7. The teacher (*begin*) _____ class at 9:00 A.M. every day.
8. The bell (*ring*) _____ at 8:55 A.M.
9. Several students often (*come*) _____ to class late.
10. They (*have, not*) _____ a good reason for being late.

QUIZ 3 Present Progressive Tense (Charts 1-1 and 1-2)

Directions: Complete the sentences. Use the present progressive form of the verbs in parentheses.

SITUATION: *At the Beach*

Example: Many people (*spend*) __are spending__ the afternoon at the beach.

1. Many kids (*play*) _____ in the water.
2. Anna and Irena (*swim*) _____ and (*dive*) _____ in the waves.
3. My little sister (*make*) _____ a sand castle.
4. Joe (*throw*) _____ a ball, and his dog (*catch*) _____ it.
5. We (*have*) _____ a picnic.
6. Mrs. Watson (*look*) _____ for seashells.
7. Mr. Watson (*fly*) _____ his kite high in the air.
8. Everyone (*get*) _____ sunburned.

QUIZ 4 Questions and Negatives in the Present (Charts 1-1 and 1-2)

A. Directions: Complete the questions with **Is she** or **Does she**.

SITUATION: *The Visitor*

Examples: __Is she__ a tourist?

__Does she__ travel often?

1. _____ have a visa?
2. _____ speak English?
3. _____ enjoying her time here?
4. _____ want to visit the capital city?
5. _____ happy here?

B. Directions: Make the second sentence in each pair a negative.

Example: It's too hot today.

It __isn't__ too cold today.

1. My parents call me every week.

   My parents _____ me every day.

2. Paolo is studying law at Oxford University.

   Paolo _____ history.

(continued on next page)

3. I remember my English teacher's name.

   I _____ my math teacher's name.

4. Jonas and Antoine are taking the bus to school.

   Jonas and Antoine _____ a taxi to school.

5. Mikael lives in St. Petersburg.

   Mikael _____ in Moscow.

QUIZ 5 Singular or Plural   (Chart 1-3)

Directions: Look at each word in *italics*. Is it a noun or a verb? Is it singular or plural? The first one is done for you.

SITUATION: *Summer*

sentence	noun	verb	singular	plural
1. The sun **shines**.		✓	✓	
2. **Flowers** grow in the garden.				
3. **Birds** sing.				
4. A boy **plays** basketball at the park.				
5. **Girls** ride bicycles.				
6. My friend **swims** every day.				
7. **Farmers** sell fresh vegetables.				
8. A man **cooks** outside on a grill.				
9. The weather **stays** warm.				
10. Everyone **wears** sunglasses.				
11. **Kids** do not have school.				

QUIZ 6 Frequency Adverbs (Chart 1-4)

A. Directions: Complete the sentences. Use appropriate frequency adverbs from the box. More than one answer may be correct.

Example: Samantha eats out three times a week.

Samantha ___often eats out___.

| always | often | sometimes | rarely | never |

1. Ari is late for work at least twice a week.

   Ari _____ late for work.

2. Every time I buy something, I use a credit card. I don't pay cash.

   When I buy something, I _____ cash.

3. About twice a month, Pat forgets his cell phone.

   Pat _____ forgets his cell phone.

4. I go running at 5:00 every morning—even when it's raining.

   I _____ running in the morning.

5. I see the dentist about once every two years.

   I _____ the dentist.

B. Directions: Add the given adverb to each sentence. Make any necessary changes to the sentence.

Example: My dad goes to work at 5:30 A.M. (*always*)

   My dad always goes to work at 5:30 A.M.

1. Does Alex go bowling? (*ever*)

   _____

2. We go to the theater more than once a month. (*seldom*)

   _____

3. Abdul is hungry at dinnertime. (*usually*)

   _____

4. I stay out past midnight on weekends. (*occasionally*)

   _____

5. Lee doesn't do his homework. (*always*)

   _____

QUIZ 7 Simple Present vs. Present Progressive (Charts 1-1 → 1-5)

Directions: Circle the correct completions.

Example: Maria sleep / (sleeps) / is sleeping until 10:00 every morning.

1. Right now, Janice **relax / relaxes / is relaxing** on her sofa.
   She **listen / listens / is listening** to music.

2. My bus usually **come / comes / is coming** at 7:50 A.M.
   It often **arrive / arrives / is arriving** a little late because of busy morning traffic.

3. A: Your phone **ring / rings / is ringing**.
   B: I'll get it.
   A: Who **call / calls / is calling**?

4. The Wilsons **clean / cleans / are cleaning** up their vegetable garden right now.
   Sometimes they **have / has / are having** a lot of work to do in the garden.

5. Frank **don't work / doesn't work / isn't working** right now. Every day,
   he **look / looks / is looking** online for a new job.

QUIZ 8 Verbs Not Usually Used in the Progressive (Chart 1-6)

Directions: Circle the correct completions. The first one is done for you.

1. Sara is busy. She **tries / (is trying)** to learn a new song on the piano.
2. I **am looking / look** for a good website about 1950s American cars.
3. I'm sorry. I **don't remember / am not remembering** your name.
4. **Do you have / Are you having** a good time?
5. I **think / am thinking** about my family now.
6. This package **belongs / is belonging** to Mr. Johnson.
7. David **doesn't have / isn't having** enough money for lunch.
8. I **think / am thinking** that my parents are wonderful.
9. Martha **looks / is looking** tired.
10. The football players **practice / are practicing** a lot this afternoon.
    They **need / are needing** a break.

QUIZ 9 Simple Present vs. Present Progressive (Charts 1-1 → 1-6)

Directions: Complete the sentences with the simple present or present progressive form of the verbs in parentheses.

Examples: Shhh. Grandma (*sleep*) ___is sleeping___ on the couch.

She (*need*) ___needs___ some rest.

1. Right now, Professor Kim (*answer*) _____ students' questions about his lecture. He usually (*do*) _____ this during the last ten minutes of each class.

2. In addition to Japanese, Tomo (*speak*) _____ four other languages. Right now, he (*speak*) _____ to his friends in English. They (*want*) _____ more practice in spoken English.

3. Look! That small plane (*fly*) _____ very low to the ground.

4. Vivian usually (*ride*) _____ her bike after work. Tonight she (*work*) _____ late. She (*ride, not*) _____ her bike.

5. I love to travel! (*you, like*) _____ to travel?

QUIZ 10 Present Verbs: Short Answers to Yes / No Questions (Chart 1-7)

Directions: Complete the conversation. Use the simple present or present progressive form of the verbs in parentheses. Give short answers to the questions as necessary.

SITUATION: *Talking about Pets*

Examples: A: (*Karl, have*) ___Does Karl have___ a cat?

B: No, ___he doesn't___. He (*be*) ___is___ allergic to cats.

A: (*your neighbors, have*) _____ a dog?

B: Yes, _____. They (*have*) _____ a bulldog.

A: (*it, be*) _____ friendly?

B: Yes, _____. It really (*like*) _____ people.

A: (*they, walk*) _____ in the park right now?

B: Yes, _____. The dog (*love*) _____ the park.

A: (*you, want*) _____ to get a dog?

B: No, _____.

QUIZ 11 Chapter Review

Directions: Circle the correct completions.

Example: A large car **(uses)/ is using** more gasoline than a small car.

1. On Sundays, we usually **go / are going** for a drive in the country. It **does / is** very relaxing.

2. A: Your homework assignment **looks / is looking** long. **Does / Is** it hard?

   B: Yes, it **does / is**. Please be quiet. I **try / am trying** to study.

3. This party is a lot of fun! We **have / are having** a wonderful time.

4. A: **Do / Are** you need anything at the store? I **leave / am leaving** now.

   B: No, I **don't / am not**. Thanks anyway.

5. My parents **believe / are believing** in hard work. They **own / are owning** two restaurants. They **work / are working** seven days a week.

6. A: What **do you do / are you doing** right now?

   B: I **send / am sending** an email to my teacher.

   A: **Do you send / Are you sending** him emails often?

   B: No, I **don't / am not**. Only when I **have / am having** to hand my homework in late.

7. When it **snows / is snowing** a lot, we can't drive because the roads **become / are becoming** too dangerous.

QUIZ 12 Chapter Review

Directions: Complete the sentences with the simple present or present progressive form of the verbs in parentheses. Add the frequency adverb where needed.

Example: Right now, Mike (*play*) __is playing__ a game on his phone. He (*play, often*) __often plays__ games.

1. My work day (*begin, usually*) _____ at 8:00 A.M. and (*end*) _____ at 5:00 P.M.

2. Listen. The birds in the trees (*sing*) _____. It's beautiful!

3. A: The baby (*cry*) _____.

   B: I know. She (*be, always*) _____ hungry.

4. A: (*you, cook*) _____ dinner now?

   B: Yes, I _____. It's almost ready.

5. Ellen (*catch*) _____ the train every evening at 6:00. Then she (*walk*) _____ from the station to her house.

6. A: (*you and Sam, want*) _____ to watch a movie later?

   B: No, we _____. We want to watch the soccer game.

7. The children (*paint*) _____ pictures right now. They are using blue, yellow, and red paint.

8. A: Why (*be*) _____ the window open?

   B: I (*know, not*) _____. It (*be*) _____ too cold!

QUIZ 13 Chapter Review

Directions: Correct the errors.

Example: My mother ~~is calling~~ *calls* me every day.

1. Rudy want to go skateboarding with his friends.

2. My sister is having four children: two girls and two boys. They are so cute!

3. Stephanie and Sarah meets at the coffee shop every Thursday night.

4. Does it rain right now?

5. I like to play tennis, but I are not a very good tennis player.

6. A: Do you have a laptop computer?

   B: No, I doesn't.

7. Nora doesn't has time to watch TV. She has too much homework.

8. Look out! That car goes too fast!

9. A: Yuko, what do you doing?

   B: I am checking my email.

10. I am practice English in class every day.

CHAPTER 1 – TEST 1

Part A *Directions:* Circle the correct completions.

1. Hurry! The bus **comes / is coming**. I **want / am wanting** to get a seat.

2. I **don't understand / am not understanding** Mrs. Brown. She never **lets / is letting** her children play outside.

3. A: Where are the children?
   B: Upstairs. They **watch / are watching** a video.

4. A: Excuse me. Is this bag yours?
   B: No, it **doesn't belong / isn't belonging** to me.

5. A: Let's stop at an ATM. I **need / am needing** some money.
   B: OK. There **is / are** an ATM on the corner. You can get some cash.

Part B *Directions:* Complete the sentences with the simple present or present progressive form of the verbs in parentheses. Add the frequency adverb where needed.

1. Look outside. The sky (*get*) _____ dark. A storm (*come*) _____ .

2. Every afternoon after school, Yasuko (*practice*) _____ the violin for several hours.

3. John (*go*) _____ to the park every day. He (*play, always*) _____ soccer with his friends. Right now, two soccer players (*kick*) _____ the ball on the field.

4. Beth (*work*) _____ at a golf course two days a week. She (*teach, often*) _____ golf classes to young children. The children (*like*) _____ her because she (*be*) _____ very patient.

5. Abdul (*have*) _____ two summer jobs. He (*pick*) _____ apples in the morning. It (*be, not*) _____ easy work. In the afternoons, he and his friends (*repair*) _____ old cars. Today, they (*work*) _____ on a 1970 Ford Mustang.

Part C *Directions:* Make the second sentence in each pair a negative.

1. Mr. and Mrs. Billings like to dance the salsa.

    Mr. and Mrs. Billings _____ to dance the tango.

2. Jeremiah wants a chocolate milkshake.

    Jeremiah _____ a strawberry milkshake.

3. My sister is a college basketball coach.

    She _____ a soccer coach.

4. The doctor sees forty patients a day.

    The doctor _____ 150 patients a day.

5. I am a taxi driver.

    I _____ a bus driver.

Part D *Directions:* Correct the verb errors.

1. The teacher never is yelling at her students. She is very patient.

2. What time are you leave school every day?

3. The Smiths no have a car. They take the bus everywhere.

4. Is Jonathan own an apartment or a house?

5. Wait. The sandwiches is ready, but not the pizza.

6. Michelle have a beautiful engagement ring from her boyfriend.

7. Hey, Kerry. Want you something to drink?

CHAPTER 1 – TEST 2

Part A *Directions:* Circle the correct completions.

1. Ruth **cuts / is cutting** her children's hair once a month.
   She **does / is doing** a good job.

2. Slow down! I **see / am seeing** some flashing lights.
   A fire truck **comes / is coming**.

3. Martha **reads / is reading** the news online every morning.
   She **doesn't want / isn't wanting** a lot of newspapers in her house.

4. My sister usually **goes / is going** to the supermarket on Fridays.

5. This chocolate cake **smells / is smelling** delicious!

Part B *Directions:* Complete the sentences with the simple present or present progressive form of the verb in parentheses.

1. My plants look dry. They (*need*) _____ water.

2. Julie is very talented. She (*play*) _____ several musical instruments well.

3. The classroom (*be*) _____ very quiet right now. Some students (*work*) _____ at their desks. Others (*write*) _____ on the board. Their teacher (*correct*) _____ some papers.

4. My aunt and uncle (*own*) _____ a small farm. Every morning they (*wake*) _____ up before sunrise. My aunt (*feed*) _____ the animals. My uncle (*take*) _____ care of the vegetable garden. They also (*have*) _____ jobs in town. My uncle (*leave*) _____ for work at 8:00 A.M. My aunt (*catch*) _____ the bus at 9:00 A.M. Today is a holiday, so they (*stay*) _____ home and (*enjoy*) _____ a quiet day off.

Part C *Directions:* Make the second sentence in each pair a negative.

1. I have two cats, a bird, and a goldfish.

   I _____ a dog.

2. The story is about three little pigs and a big bad wolf.

   The story _____ about three bears and a little girl.

3. Henry and Marie enjoy eating at expensive restaurants.

   Henry and Marie _____ eating at fast-food restaurants.

4. The coffee tastes strong.

   The coffee _____ weak.

5. We are going to Yellowstone Park.

   We _____ to the Grand Canyon.

Part D *Directions:* Correct the verb errors.

1. Are you always going to school by bus?

2. I no like movies with sad endings.

3. Oh no, look! A rat plays in the garbage can.

4. Is Maria work on Saturdays?

5. The books is on sale, but not the magazines.

6. Mr. Green is almost 80, but he isn't want to live with his children.

7. Ted eats Mexican food often. He is loving Mexican food.

CHAPTER 2　Past Time

QUIZ 1　Simple Past Tense　(Charts 2-1 → 2-3)

Directions: Change the sentences to past time. Use the simple past and *yesterday* or *last*.

Example: Sue exercises every day at 8:00.

Sue ___exercised yesterday at 8:00___.

I see my cousins every Sunday.

___I saw my cousins last Sunday___.

1. Andrew works for eight hours every day.

   Andrew _____.

2. Mark and Jan go to bed at 10:00 every night.

   Mark and Jan _____.

3. The alarm on my phone rings at 6:00 every morning.

   The alarm on my phone _____.

4. My grandparents visit us every month.

   My grandparents _____.

5. I take a nap every afternoon.

   I _____.

6. Dr. Hughes teaches medical students every Tuesday evening.

   Dr. Hughes _____.

7. Victoria buys coffee at Cory's Coffee Shop every morning.

   Victoria _____.

8. Mr. Wilson walks to the farmers' market every Saturday.

   Mr. Wilson _____.

9. Anne calls her best friend every week.

   Anne _____.

10. It rains in Seattle every winter.

    It _____.

QUIZ 2 Simple Past: Negatives (Charts 2-1 → 2-3)

Directions: The statements have incorrect information. Make true statements by using the negative.

Example: Dinosaurs lived in the ocean.

Dinosaurs ____*didn't live*____ in the ocean. They lived on land.

1. It was 100 degrees Celsius yesterday. It snowed.

    It was 100 degrees Celsius yesterday. It _____. It was sunny.

2. Marco Polo traveled in South America.

    Marco Polo _____ in South America. He traveled in Asia.

3. American astronaut Scott Kelly went to the moon.

    Scott Kelly _____ to the moon. He went to the International Space Station.

4. The Beatles' first U.S. tour was in 1963.

    The Beatles' first U.S. tour _____ in 1963. It was in 1965.

5. Mozart wrote plays.

    Mozart _____ plays. He composed music.

6. Nelson Mandela became president of South Africa in 2014.

    Nelson Mandela _____ president in 2014. He became president in 1994.

7. People drove cars in the 15th century.

    People _____ cars in the fifteenth century. They rode horses.

8. The 2016 Summer Olympics were in Sochi, Russia.

    The 2016 Summer Olympics _____ in Sochi. They were in Rio de Janeiro, Brazil.

9. An earthquake damaged New Orleans in 2005.

    An earthquake _____ New Orleans in 2005. A hurricane did.

10. In *Alice in Wonderland*, Alice met three bears.

    Alice _____ three bears. She met a white rabbit.

QUIZ 3 Simple Past: Questions (Charts 2-1 → 2-3)

Directions: Complete the questions with **Did, Was,** or **Were.**

Examples: _____Did_____ you ride the bus to work today?

_____Was_____ the bus on time?

SITUATION: *At the Supermarket*

1. _____ you go to the supermarket yesterday?
2. _____ the vegetables fresh?
3. _____ the meat expensive?
4. _____ the clerk helpful?
5. _____ you pay with cash or a credit card?

SITUATION: *A Weekend at Home*

6. _____ you stay home last weekend?
7. _____ you and your family tired?
8. _____ anyone take a nap?
9. _____ you do any housework?
10. _____ you bored?

QUIZ 4 Simple Past: Negatives and Questions (Charts 2-1 → 2-3)

Directions: Write the question and negative forms for each sentence.

Example: The plane arrived on time.

QUESTION: ___Did the plane arrive___ on time?

NEGATIVE: ___The plane didn't arrive___ on time.

1. Nina used her laptop in class.

   QUESTION: _____ her laptop computer in class?

   NEGATIVE: _____ her laptop computer in class.

2. The restaurant was expensive.

   QUESTION: _____ expensive?

   NEGATIVE: _____ expensive.

(continued on next page)

3. Mrs. Douglas knocked on the door.

   QUESTION: _____ on the door?

   NEGATIVE: _____ on the door.

4. Mark took photos with his phone.

   QUESTION: _____ photos with his phone?

   NEGATIVE: _____ photos with his phone.

5. Julie got a new job.

   QUESTION: _____ a new job?

   NEGATIVE: _____ a new job.

QUIZ 5 Simple Past: Questions and Answers (Charts 2-1 → 2-3)

Directions: Complete the conversations with the correct past tense form of the words in parentheses. Add short answers to questions as necessary.

Example: A: (*Beethoven, write*) ___Did Beethoven write___ symphonies?

B: Yes, _he did_. He (*write*) __wrote__ nine symphonies.

1. A: (*Sara, eat*) _____ salad for lunch yesterday?

   B: Yes, _____. She (*eat*) _____ a spinach and tomato salad.

2. A: (*you, go*) _____ to work last week?

   B: No, _____. I (*be*) _____ on vacation.

3. A: (*Fred, go*) _____ on a long hike last weekend?

   B: Yes, _____. He (*hike*) _____ over ten miles.

4. A: (*John and Linda, travel*) _____ to Brazil in 2018?

   B: Yes, _____. They (*spend*) _____ two weeks in Recife.

5. A: (*you, sleep*) _____ well last night?

   B: No, _____. I (*hear*) _____ noises outside and

   (*stay*) _____ awake for a long time.

6. A: (*the monster movie, scare*) _____ the children?

   B: No, _____. It (*scare, not*) _____ them.

   It (*make*) _____ them laugh.

QUIZ 6 Spelling of -ing and -ed Forms (Charts 2-5)

Directions: Write the **-ed** and the **-ing** forms of the verbs. Pay attention to spelling. The first one is done for you.

	-ed	-ing
1. want	wanted	wanting
2. love		
3. help		
4. visit		
5. worry		
6. hop		
7. play		
8. smile		
9. tie		
10. stop		
11. study		
12. rain		
13. prefer		
14. fix		
15. happen		
16. enjoy		

QUIZ 7 Simple Past (Charts 2-1 → 2-5)

Directions: Write the simple past form of the verbs in *italics*. The first one is done for you.

1. I *want* soup for lunch. _____wanted_____
2. I *run* at night. _____
3. Jay *is* a chef. _____
4. This GPS *works* well. _____
5. The passports *aren't* ready. _____
6. You *understand* the problem. _____
7. Antonio *ties* his shoes. _____
8. Julia *has* many friends. _____
9. Ed *tries* to help. _____
10. Our teacher *doesn't yell* at us. _____
11. Your book *isn't* here. _____
12. My tooth *hurts*. _____
13. You *know* my cousins. _____
14. My dad *leaves* at 7:10. _____
15. Ruth *cleans* her room. _____
16. They *don't have* time. _____

QUIZ 8 Simple Past (Charts 2-1 → 2-5)

Directions: Complete the sentences with the simple past form of the verbs in parentheses.

Example: Mari (*work*) _____worked_____ six days in a row last week.

1. Abdou (*send*) _____ his parents three emails yesterday.

2. I (*take*) _____ some medicine for my headache this morning.

3. You (*bring*) _____ home lots of bags and packages. What (*you, buy*) _____ at the store?

4. The ground (*shake*) _____ several times after the big earthquake.

5. I (*pick*) _____ up lots of books at the sale last week.

6. Jeff (*ask*) _____ a question during the meeting, but his boss (*hear, not*) _____ him.

7. I (*order*) _____ grilled fish at the restaurant. It (*be*) _____ delicious.

QUIZ 9 Simple Past (Charts 2-1 → 2-5)

Directions: Complete the paragraph with the simple past form of the verbs in parentheses. The first one is done for you.

SITUATION: *Happy Birthday to Me!*

I (*have*) ___had___₁ a wonderful birthday last Wednesday. It (*be*) _____₂ my 30th birthday. I (*want, not*) _____₃ a party, so I (*tell, not*) _____₄ my friends about it. I (*take*) _____₅ the day off from work. I (*wake*) _____₆ up early and (*decide*) _____₇ to go shopping. I (*ride*) _____₈ the bus downtown and (*get*) _____₉ off at Camden Market. It (*be, not*) _____₁₀ crowded because it (*be*) _____₁₁ the middle of a weekday. I (*shop*) _____₁₂ for clothes and (*buy*) _____₁₃ a leather jacket and some boots. I (*find*) _____₁₄ a restaurant with a view of the city and (*order*) _____₁₅ a delicious lunch. After that, I (*walk*) _____₁₆ to a nearby café and (*have*) _____₁₇ dessert and coffee. Finally, I (*sit*) _____₁₈ in a small park and (*enjoy*) _____₁₉ the beautiful weather. At the end of the day, I (*feel*) _____₂₀ relaxed and happy. It (*be*) _____₂₁ one of my best birthdays ever.

QUIZ 10 The Past Progressive (Chart 2-6)

Directions: Complete the sentences and questions with the past progressive form of the verbs in parentheses.

Example: Joe (*sleep*) __was sleeping__ at 11:00 last night.

1. Between 5:00 and 6:00 P.M. yesterday, Harry (*exercise*) _____ at the gym.

2. My parents and I (*visit*) _____ my grandparents at this time last year.

3. What (*do*) _____ you _____ at noon today? I didn't see you in the cafeteria.

4. I (*live, not*) _____ in Florida in 2015. I (*study*) _____ in New York.

5. Lucy and Barbara (*attend*) _____ a conference in San Francisco all last week.

6. When I was 13, I (*take*) _____ ballet lessons and (*learn*) _____ to play the guitar.

7. A: Mr. Davis was out of the office for three days last week. (*travel*) _____ he _____ on business?

   B: Yes, he _____. He went to Dallas.

QUIZ 11 Understanding Past Time (Charts 2-1 → 2-7)

A. *Directions:* Match the numbered sentences with the lettered sentences below that express the same idea. Write the letter on the line. The first one is done for you.

1. I ended the call. My doorbell was ringing. __c__

2. I ended the call. My doorbell rang. ____

3. I woke up. My mother smiled at me. ____

4. I woke up. My mother was smiling at me. ____

5. I looked out the window. A bird flew by. ____

6. I looked out the window. A bird was flying by. ____

   a. First, the bird began to fly. Then I looked out the window.
   b. First, I woke up. Then my mother smiled.
   c. First, my doorbell started to ring. Then I ended the call.
   d. First, I looked out the window. Then the bird started to fly.
   e. First, my mother began to smile. Then I woke up.
   f. First, I ended the call. Then my doorbell began to ring.

(continued on next page)

B. Directions: Read the sentences and answer the questions.

Example: Jordan was driving his car. He was honking his horn.
Lisa was driving her car. She honked her horn.
QUESTION: *Which driver honked more?*   (Jordan)   Lisa

1. Juliette was driving to the bank. Her mother called.
   Abby's mother called. She drove to the bank.
   QUESTION: *Who got a call in the car?*   Juliette   Abby

2. When dinner ended, Peter sent a text to his friend.
   When dinner ended, Max was sending a text to his friend.
   QUESTION: *Who sent a text after dinner ended?*   Peter   Max

3. The boys were playing soccer. The rain stopped.
   The rain stopped. The girls played soccer.
   QUESTION: *Who played soccer in the rain?*   the boys   the girls

4. Karen was checking her email during breakfast.
   Marek checked his email during breakfast.
   QUESTION: *Who probably spent more time checking email?*   Karen   Marek

5. Faisal was watching a movie. He was eating dinner.
   Atsushi watched a movie. He ate dinner.
   QUESTION: *Who ate dinner during the movie?*   Faisal   Atsushi

QUIZ 12 Simple Past vs. Past Progressive (Charts 2-1 → 2-7)

Directions: Complete each sentence using the information in the chart. Use the simple past for one clause and the past progressive for the other. The first one is done for you. More than one answer may be correct.

SITUATION: *Simone, Linda, and Janet are busy teachers.*

Activity in Progress	Simone	Linda	Janet
work in her office	answer the phone	meet with students	check her email
wait for the bus	send a text message	call her son	read a novel
sit at the snack bar	drink a soda	meet an old friend	eat a sandwich

1. While Simone _____*was working in her office, she answered the phone*_____.
2. Linda _____ while _____.
3. While Janet _____.
4. Simone _____ while _____.
5. While Linda _____.
6. Janet _____ while _____.

QUIZ 13 Simple Past and Past Progressive (Charts 2-1 → 2-7)

Directions: Complete the sentences with the simple past or past progressive form of the verbs in parentheses. Pay attention to irregular past verb forms.

Example: I'm sorry I'm late. I (*hear, not*) ____didn't hear____ the alarm. It

(*ring*) ____was ringing____ too quietly!

1. Last week, my sister and her husband (*decide*) _____ to sell their home and move into an apartment.

2. The Norlings (*spend*) _____ last winter in Mexico. While they (*live*) _____ there, they (*study*) _____ Spanish.

3. Meg (*make*) _____ her chocolate fudge cake for the baking contest, and she (*win*) _____ first prize.

4. George (*plan*) _____ a retirement party for his office manager, and more than 100 people showed up.

5. Melissa's dog (*run*) _____ away from home while she (*travel*) _____ in South America last month. She (*try*) _____ to find him when she (*return*) _____, but she (*have, not*) _____ any luck.

6. Last Tuesday, we (*have*) _____ a terrible windstorm. The wind (*blow*) _____ and (*howl*) _____ all night long. While I (*try*) _____ to sleep, my cats (*become*) _____ frightened and (*hide*) _____ under the bed.

7. Geoff (*live*) _____ in Paris for three months last year. He (*speak*) _____ French every day.

QUIZ 14 Simple Past and Past Progressive (Charts 2-1 → 2-7)

Directions: Complete the paragraph with the simple past or the past progressive form of the verbs in parentheses. The first sentence is done for you.

SITUATION: *Dinner in a Foreign Language*

Yesterday, while Sara (*walk*) __was walking__(1) home from work, she (*see*) __saw__(2) an old friend. While they (*talk*) _____(3), another friend of Sara's (*join*) _____(4) them. They all (*chat*) _____(5) for a while and then (*decide*) _____(6) to have dinner at a new restaurant nearby. At the restaurant, they (*know, not*) _____(7) what to order because the menu (*be*) _____(8) in another language. They (*ask*) _____(9) the waiter for suggestions, but he (*be, not*) _____(10) very helpful. They finally (*point*) _____(11) to some food on the menu and (*hope*) _____(12) for a tasty meal.

QUIZ 15 Expressing Past Time: Understanding Time Clauses (Chart 2-8)

Directions: Write "1" before the action that started first and "2" before the action that started second. The first one is done for you.

SITUATION: *Joann Baked a Cake*

1. Before Joann baked the cake, she found a recipe in her cookbook.

   _2_ Joann baked the cake.

   _1_ She found a recipe in her cookbook.

2. After she found a recipe she liked, Joann made a shopping list.

   ___ Joann found a recipe she liked.

   ___ She made a shopping list.

3. Joann shopped until she found everything on her list.

   ___ Joann shopped.

   ___ She found everything on her list.

(continued on next page)

4. When Joann finished shopping, she went home to bake the cake.

   ____ Joann finished shopping.

   ____ She went home to bake the cake.

5. Joann turned on the oven as soon as she got home.

   ____ Joann turned on the oven.

   ____ She got home.

6. She measured flour, sugar, and salt before she stirred the eggs and milk.

   ____ She measured flour, sugar, and salt.

   ____ She stirred the eggs and milk.

7. She poured the batter into the pan as soon as she mixed everything together.

   ____ She poured the batter into the pan.

   ____ She mixed everything together.

8. Joann took the cake out of the oven when it finished baking.

   ____ Joann took the cake out of the oven.

   ____ The cake finished baking.

9. Joann's family waited patiently until she cut the cake.

   ____ Joann's family waited patiently.

   ____ Joann cut the cake.

10. Joann's family began to eat as soon as she served the cake.

    ____ Joann's family began to eat.

    ____ Joann served the cake.

11. After they enjoyed the cake, they thanked Joann for the delicious dessert.

    ____ They enjoyed the cake.

    ____ They thanked Joann for the delicious dessert.

QUIZ 16 Expressing Past Time: Using Time Clauses (Chart 2-8)

Directions: Combine the two sentences into one sentence by using time clauses. Add punctuation as necessary.

Example: First: Patrick lost a library book. *Then:* He paid for it.

After _____*Patrick lost a library book, he paid for it*_____.

1. *First:* Donna got a ticket for speeding. *Then:* She drove home very slowly.

    After _____.

2. *First:* Eric got home. *Then:* He took a shower.

    _____ as soon as _____.

3. *First:* Rick was cooking dinner. *Then:* He burned his hand.

    _____ while _____.

4. *First:* Kevin did his homework. *Then:* He watched a movie.

    Before _____.

5. *First:* My family lived in Germany. *Then:* My dad got a job in Italy.

    Until _____.

6. *First:* I put on my pajamas. *Then:* I went to bed.

    As soon as _____.

7. *First:* The band started to play. *Then:* Everyone started dancing.

    _____ after _____.

8. *First:* Joy's parents drove her to school. *Then:* She got her driver's license.

    _____ before _____.

9. *First:* The students were nervous. *Then:* The exam began.

    _____ until _____.

10. *At the same time:* Maria and Lucio were exercising. Their children were playing a board game.

    While _____.

QUIZ 17 Expressing Past Habit: *Used To* (Chart 2-9)

Directions: Complete each sentence with **used to** and a verb from the box. The first one is done for you.

be	drink	have	ski	swim	work
bark	eat	live	speak	wake up	

1. I ____used to eat____ peanuts as a snack, but now I'm allergic to peanuts.
2. I _____ in the ocean when I was a child, but now the water is too cold for me.
3. My hair _____ long, but now I have short hair.
4. I _____ a lot of carrot juice, but then my skin began to turn orange.
5. My father _____ 12 hours a day until he had a heart attack.
6. You _____ a lot of homework when you were in high school.
7. The baby _____ several times during the night, but now she sleeps through the night.
8. Mark and Marissa _____ in Chicago, but now they live in Omaha.
9. When I lived in Montana, I _____ every winter. There was a lot of snow.
10. Susan _____ Spanish, but now she doesn't remember much.
11. Our dog _____ at cars, but now she just watches them.

QUIZ 18 Chapter Review

Directions: Correct the errors.

1. Joe no walk to work yesterday. He take the bus.
2. Mary was go to the emergency room at midnight last night.
3. While Dr. Dann listen to his patient, his cell phone rang. He wasn't answer it.
4. Marco use to swim every week, but now he prefers working out at the gym.
5. After Evie was getting a new job, she celebrates her success with her friends.
6. When the phone was ringing at 11:00 last night, I was in a deep sleep. I almost not hear it.

QUIZ 19 Chapters 1 and 2 Review

Directions: Complete the sentences using the words in parentheses. Use the simple present, present progressive, simple past, or past progressive form of the verbs.

Example: Jack usually (*pay*) _____*pays*_____ his telephone bill with a credit card, but last month he (*pay*) _____*paid*_____ by check.

1. A: What time (*you, get up*) _____ every day?

   B: I usually (*wake up*) _____ before the sun

   (*rise*) _____, but yesterday I (*sleep*) _____

   until 9:00 A.M. It (*feel*) _____ wonderful!

2. At least once a week, the cat (*catch*) _____ a mouse, but last week she

   (*catch*) _____ three mice. While she (*chase*) _____ one

   of them, she (*knock*) _____ over my favorite plant.

3. Last Saturday, before I (*leave*) _____ for the beach, I

   (*buy*) _____ gas for my motorbike.

4. A: Arnie, why are you on the floor? What (*you, do*) _____?

   B: I (*clean*) _____ the kitchen floor. Your dog (*get*) _____

   into the garbage last night while we (*sleep*) _____. He

   (*make*) _____ a mess!

5. The Smiths often travel to Africa on vacation, but last year they (*travel*)

   _____ to Southeast Asia. While they (*travel*) _____,

   they (*visit*) _____ several small villages. They

   (*stay, not*) _____ in any big cities. They (*enjoy*) _____

   visiting a new continent.

6. The Bentons usually (*watch*) _____ movies on Saturday night, but

   yesterday they (*watch*) _____ their neighbor's children instead.

7. When Jack (*step*) _____ off the train, his wife (*wait*) _____

   for him on the station platform. He (*be*) _____ so happy to see her.

8. We used to (*live*) _____ in Mumbai, but now we (*live*)

   _____ in Delhi. After we (*move*) _____ here, I

   (*miss*) _____ my friends and school in Mumbai very much. Now I

   (*be*) _____ used to Delhi, and I (*like*) _____ living here.

9. A: (*you, read*) _____ the news this morning?

   B: No, I _____. Why?

   A: I saw an article about your former boss. He (*go*) _____ to jail for

   theft last week.

CHAPTER 2 – TEST 1

Part A ***Directions:*** Complete the paragraph with the simple past or past progressive form of the verbs in parentheses.

SITUATION: *Unexpected Adventure*

Last weekend, we (*take*) _____1_____ a day trip on our sailboat. We (*see*) _____2_____ a small island in the middle of a lake. We (*want*) _____3_____ to explore it, so we (*leave*) _____4_____ our boat on the beach on the island. We (*find*) _____5_____ a trail and (*hike*) _____6_____ around the island. While we (*hike*) _____7_____, we (*hear*) _____8_____ some strange noises in the bushes. We (*stand*) _____9_____ on the trail and (*wait*) _____10_____. While we (*wait*) _____11_____, a small bear (*come*) _____12_____ out of the woods. It (*eat*) _____13_____ blackberries. After we (*see*) _____14_____ the bear, we slowly (*walk*) _____15_____ backwards. Fortunately, the bear (*follow, not*) _____16_____ us, and we (*get*) _____17_____ back to our boat safely.

Part B ***Directions:*** Complete the conversation with the correct present or past form of the verbs in parentheses.

A: How old were you when you (*learn*) _____ to swim?

B: I (*be, not*) _____ very old, probably three or four. My father first (*teach*) _____ me to blow bubbles under water. Then he (*show*) _____ me how to kick properly.

A: (*you, like*) _____ to swim now?

B: I (*love*) _____ it!

Part C *Directions:* Circle the correct completions.

1. A: **Do / Did / Were** you remember to call the dentist?

   B: Yes. I **make / made / was making** an appointment for Tuesday.

2. A: What time **does / was / did** Jason come home last night?

   B: I don't know. I **don't hear / didn't hear / wasn't hearing** him.

3. A: How are you doing? You **look / looked / were looking** tired.

   B: I **am / was / did**. I **am not sleeping / wasn't sleeping / didn't sleep** well last week, and I'm still tired.

Part D *Directions:* Correct the errors.

1. Bees used to making honey in a tree next to our house until lightning hit the tree.

2. Carol doesn't go to work yesterday because her son sick.

3. Doug and Peter were having a party last weekend. Everyone from our class come.

4. Did you upset with your test results yesterday?

5. After Bill woked up, he get up.

6. Liz and Ron planed to get married last summer, but just before the wedding, Ron is losing his job. Now they wait until next summer.

CHAPTER 2 – TEST 2

Part A ***Directions:*** Complete the paragraph with the simple past or past progressive form of the verbs in parentheses.

SITUATION: *Traffic Jam*

Traffic was terrible yesterday! There (*be*) _____ an accident on the freeway. I (*sit*) _____ in my car for 15 minutes and (*go, not*) _____ anywhere. While I (*sit*) _____ in the car, I (*watch*) _____ the other drivers. One woman (*talk*) _____ on her cell phone the whole time. Another driver (*try*) _____ to calm her baby. He (*cry*) _____. An elderly man (*eat*) _____ an ice-cream cone. He (*look*) _____ like the only happy driver. Finally, the police (*come*) _____ and traffic (*begin*) _____ to move again. After we (*wait*) _____ so long, we all (*feel*) _____ thankful. I (*be*) _____ very tired when I (*get*) _____ home.

Part B ***Directions:*** Complete the conversation with the correct present or past form of the verbs in parentheses.

A: Hi, Bill. What (*you, do*) _____?

B: I (*look*) _____ at pictures of my vacation. I (*go*) _____ to my cousins' house for a week.

A: Where (*they, live*) _____?

B: They live on a lake in the country.

A: Cool! What (*you, do*) _____ there?

B: One day we (*rent*) _____ a canoe. We (*spend*) _____ all afternoon fishing and swimming. We also (*take*) _____ long walks in the forest. Once, while we (*walk*) _____, we (*see*) _____ some deer.

A: That sounds like fun!

Part C *Directions:* Circle the correct answers.

A: Where **are / were / did** you have lunch yesterday?

B: At Ellen's new restaurant. I **am having / was having / had** chicken and rice.

A: **Is / Was / Did** your meal good?

B: Yes, really good! And the owner is really friendly. While I **am eating / ate / was eating**, Ellen **comes / came / was coming** out to say hello.

Part D *Directions:* Correct the errors.

1. Dr. Martin used to working in a hospital, but now she has a private practice.
2. I was home alone last night. First, I cooked dinner. Then I was washing the dishes.
3. Matt busy yesterday. He wasn't go to the party.
4. The Millers buy a restaurant last month. They open for business last week.
5. The baby tryed to crawl a few times, but her legs aren't strong enough yet.
6. Professor Scott no have time to help us with our lab experiment yesterday. Maybe she can today.

CHAPTER 3 Future Time

QUIZ 1 Understanding Present, Past, and Future (Chapters 1, 2, and 3)

Directions: Circle the correct time word.

Example: Zach drove to work in bad traffic.   every day   **(yesterday)**

1. Niko is going to clean his apartment.   yesterday   tomorrow
2. Bill will celebrate his birthday.   every day   tomorrow
3. Do you work in the city?   every day   yesterday
4. Susan woke up early.   yesterday   tomorrow
5. Does Ming take the bus to school?   every day   yesterday
6. Sara broke her leg.   every day   yesterday
7. Will you be here?   yesterday   tomorrow
8. The mail comes at 1:00 P.M.   every day   yesterday
9. Is Alexa going to visit us?   every day   tomorrow
10. Dr. Hayashi saw two patients at the hospital.   yesterday   tomorrow

QUIZ 2 Will and Be Going To (Charts 3-1 → 3-3)

Directions: Complete the sentences with *will* and *be going to* and the verbs in parentheses.

Example: (*arrive*) The plane _____will arrive_____ soon.

The plane _____is going to arrive_____ soon.

1. (*have*) We _____ a party tomorrow.

   We _____ a party tomorrow.

2. (*meet*) Our book club _____ next weekend.

   Our book club _____ next weekend.

3. (*return*) I _____ the library books tomorrow morning.

   I _____ the library books tomorrow morning.

4. (*be*) Hurry up! You _____ late!

   Hurry up! You _____ late!

5. (*fly*) The Wilsons _____ to Toronto next Saturday.

   The Wilsons _____ to Toronto next Saturday.

QUIZ 3 Questions with *Will* and *Be Going To* (Charts 3-2 and 3-3)

Directions: Write questions with **will** and **be going to**.

Example: you / study

      _____Will you study_____ tomorrow?

      Are you going to study tomorrow?

1. the cat / catch

   _____ the mouse?

   _____ the mouse?

2. Dr. Brown / retire

   _____ next year?

   _____ next year?

3. your family / be

   _____ at the wedding?

   _____ at the wedding?

4. our team / win

   _____ the game?

   _____ the game?

5. Mr. and Mrs. Bell / find

   _____ their lost puppy?

   _____ their lost puppy?

QUIZ 4 Be Going To (Charts 3-1 and 3-2)

Directions: Complete the conversations with **be going to** and the words in parentheses.

SITUATION: *Making Plans*

Examples: A: What (*you, do*) ___are you going to do___ next Saturday?

B: I (*shop*) ___am going to shop___ for a wedding gift for my niece.

1. A: When (*we, leave*) _____ for the airport?

   B: We (*leave*) _____ in about 15 minutes.

2. A: (*you, stop*) _____ at the market after work?

   B: Yes, I (*buy*) _____ some fresh vegetables.

3. A: My mother (*visit, not*) _____ us during the holidays this year.

   B: I'm sorry to hear that. What (*she, do*) _____?

   A: She (*spend*) _____ the holidays in Florida.

4. A: Josh and Joy (*go*) _____ to a jazz concert next weekend.

   B: That sounds great! Who (*perform*) _____?

   A: Jazz Connection.

   B: Oh, they are really good. I'm sorry I (*be, not*) _____ there.

Future Time 35

QUIZ 5 Will (Charts 3-1 and 3-3)

Directions: Complete the conversation with **will** and the words in parentheses.

SITUATION: *Visiting Vancouver, Canada*

Example: Kevin and his dad (*travel*) ____will travel____ from Seattle to Vancouver tomorrow.

KEVIN: Dad, how long is the trip to Vancouver?

DAD: We (*drive*) _____(1)_____ for about three hours, but crossing the border into Canada (*take*) _____(2)_____ at least 30 minutes.

KEVIN: Maybe the traffic (*be, not*) _____(3)_____ too bad because it's a weekday. I (*listen*) _____(4)_____ to music to make the time pass quickly.

DAD: I think the weather in Vancouver (*be*) _____(5)_____ terrible this week. The weather report says it (*rain*) _____(6)_____ .

KEVIN: That's OK, Dad. I (*bring*) _____(7)_____ my raincoat with me.

DAD: That's a good idea. We (*take*) _____(8)_____ an umbrella, too.

KEVIN: (*go, we*) _____(9)_____ to Stanley Park while we are there?

DAD: Sorry, but we (*have, not*) _____(10)_____ enough time. Maybe we can go next time.

QUIZ 6 Contractions with Will and Be Going To (Charts 3-2 → 3-4)

Directions: Write the contractions for the words in *italics*.

SITUATION: *Lunch Time!*

Example: *I will* help you make lunch. ____I'll____

1. *We are* going to make cheese sandwiches. _____

2. First, *we will* slice the cheese. _____

3. Then *I am* going to butter the bread. _____

4. *You are* going to prepare the lettuce and tomatoes. _____

5. Clara and Kathy *are not* going to have tomatoes on their sandwiches. _____

6. Richard is very hungry. *He is* going to eat two sandwiches. _____

7. *He will* have mayonnaise and mustard on his sandwiches. _____

8. *You will* slice some fruit and open a bag of chips. _____

9. We *will not* have any dessert. _____

10. *We are* going to enjoy a delicious lunch! _____

QUIZ 7 Be Going To vs. Will (Chart 3-5)

Directions: Read the conversations. Think about the meaning of the verbs in *italics* in each conversation. Decide if the speaker is making a prediction, has a prior plan, or decides or volunteers at the moment of speaking. Circle the correct option.

Example: A: My mom *is* probably *going to call* today. (prediction) prior plan decide/volunteer
B: Tell her hello from me.

1. A: Mary has a beautiful singing voice!
   B: I think she *will be* a famous singer someday.   prediction prior plan decide/volunteer

2. A: Do you have anything special planned for the weekend?
   B: No, we*'re going to stay* home and *relax*.   prediction prior plan decide/volunteer

3. A: The sink is full of dishes.
   B: I *will wash* them for you.   prediction prior plan decide/volunteer

4. A: The sky is really dark and cloudy.
   B: I think it*'s going to rain*.   prediction prior plan decide/volunteer

5. A: It's almost midnight. I have to hurry!
   B: Your parents *will be* really angry if you are late.   prediction prior plan decide/volunteer

6. A: Where is your appointment?
   B: Downtown. I*'m going to take* the bus.   prediction prior plan decide/volunteer

7. A: The movie starts at 7:30.
   B: OK. I *will meet* you at the theater.   prediction prior plan decide/volunteer

8. A: Do you have any pets?
   B: No, but we *are going to get* a dog next month.   prediction prior plan decide/volunteer

9. A: We want to go to Spain in October.
   B: That *will be* a wonderful trip!   prediction prior plan decide/volunteer

10. A: What are we having for dinner?
    B: I guess I *will cook* some chicken and vegetables.   prediction prior plan decide/volunteer

QUIZ 8 Be Going To vs. Will (Chart 3-5)

Directions: Complete the conversations with the correct form of **be going to** or **will**.

Example: A: What are you doing this afternoon?

B: Maggie and I ___are going to___ go to a movie.

A: I ___'ll___ go with you if it is OK with you.

B: Sure!

1. A: I just spilled some milk.

   B: Don't worry. I _____ clean it up.

2. A: Why did Greg rent a truck?

   B: He _____ pick up a new bed.

3. A: What are your plans for the school break?

   B: We _____ to visit my sister in Texas.

4. A: How does this dress look?

   B: I think it's a little too short.

   A: OK. I guess I _____ buy the other one.

5. A: Do you have Joan's new phone number?

   B: Yes. Let me check my phone. I _____ find it for you.

6. A: There's a party at Scott's this weekend.

   B: I know. Tony and I _____ bring chips and soda.

7. A: This box is too heavy!

   B: Put it down. You _____ hurt your back. I _____ carry it for you.

8. A: Pat is at the hardware store. He _____ buy a new hammer.

   B: Really? Why?

   A: We _____ fix our fence. It blew down in the storm last weekend.

   B: That's too bad.

QUIZ 9 Certainty About the Future (Chart 3-6)

Directions: Decide if the speaker is 100% sure, 90% sure, or 50% sure. Put a check (✓) in the box. The first one is done for you.

SITUATION: *A New Hotel*

	100%	90%	50%
1. The new hotel is not going to open next month.	✓		
2. Maybe it will open the month after that.			
3. Many people will probably be upset about the change.			
4. The hotel may give them lower prices for future stays.			
5. The owners will refund travelers their money.			
6. Amanda isn't going to pass her classes.			
7. She won't be surprised about her poor grades.			
8. She'll probably be a little upset.			
9. Amanda's parents may be very upset.			
10. Maybe they will be strict with her.			
11. Amanda probably won't be able to go out with her friends.			

SITUATION: *Amanda's Problem*

QUIZ 10 Certainty About the Future (Chart 3-6)

Directions: Use the given words to make two predictions about each situation. Include the words in parentheses. Use **will** or **be going to** where needed. More than one answer may be correct.

Example: SITUATION: Kelly drank too much coffee.

Kelly \ not \ be able to sleep (*probably*)

  Kelly probably isn't going to be able to sleep.

she \ lie \ awake all night (*maybe*)

  Maybe she will lie awake all night.

1. SITUATION: Tom's car is making a strange noise.

   Tom \ take \ the car to the mechanic's tomorrow (*may*)

   _____

   the mechanic \ fix \ the problem (*probably*)

   _____

2. SITUATION: Max doesn't like his boss.

   Max \ quit \ his job (*maybe*)

   _____

   Max's boss \ be \ not \ happy about that (*probably*)

   _____

(continued on next page)

3. **Situation:** Yujung speaks Korean, Chinese, and English.

   Yujung \ get \ a job in international business (*may*)

   _____

   she \ earn \ a good salary with her language skills (*maybe*)

   _____

4. **Situation:** The Shepherds are planning to paint their house.

   The Shepherds \ buy \ some paint \ next weekend (*may*)

   _____

   they \ paint \ not \ their house \ a dark color (*probably*)

   _____

5. **Situation:** My children enjoy watching movies.

   My children \ go \ not \ to a movie theater this weekend (*probably*)

   _____

   they \ watch \ a movie at home (*maybe*)

   _____

QUIZ 11 Expressing the Future in Time Clauses and *If*-Clauses   (Chart 3-7)

Directions: Use the given information to make sentences with a future meaning. Use ***will*** for the future.

Example: buy a new car \ earn enough money

   As soon as Matt __earns enough money__, he __will buy a new car__.

1. pick up their airplane tickets \ fly to Thailand

   Before the Smiths _____, they _____.

2. get dressed \ go to work

   As soon as Sonya _____, she _____.

3. feel better \ stay home

   Chris _____ until he _____.

4. get a driver's license \ take the driving test

   After Mr. Hill _____, he _____.

5. check her answers \ turn in her test paper

   Beatriz _____ before she _____.

(continued on next page)

6. get a new phone \ text me

   When Thomas _____, he _____.

7. win a lot of money \ quit her job

   If Janice _____, she _____.

8. go to the staff meeting \ go home

   David and Sam _____ after they _____.

9. wash her hands \ make lunch

   Before Ellen _____, she _____.

10. be in bed by midnight \ finish his homework

    Josh _____ if he _____.

QUIZ 12 Using *Be Going To* and the Present Progressive to Express Future Time (Chart 3-8)

A. **Directions:** Rewrite the sentences. Use the present progressive to express the future.

   Example: The restaurant is going to close for one week.

   The restaurant ___*is closing*___ for one week.

1. I am going to change schools next year.

   I _____ schools next year.

2. The Andersons are going to have a barbecue on Saturday.

   The Andersons _____ a barbecue on Saturday.

3. Liz is going to join a hiking club this summer.

   Liz _____ a hiking club this summer.

4. Dr. Allen is going to take time off from work in April.

   Dr. Allen _____ time off from work in April.

5. Are you going to move to New Jersey soon?

   _____ you _____ to New Jersey soon?

B. **Directions:** Circle all the correct time expressions to complete each sentence.

   Example: The Moores are staying home (tomorrow.) (tonight.) yesterday. (now.)

   1. Emil is cooking dinner   tonight.   right now.   every day.   all next week.
   2. Nina is going to graduate   today.   tomorrow.   tonight.   next semester.   every day.
   3. The clinic is opening   today.   in one hour.   next week.   every day.   next year.
   4. I am going to work   now.   in ten minutes. tomorrow.   next week.
   5. I am going to go to work   tomorrow. last week.   soon.   in a few minutes.

QUIZ 13 Using the Simple Present to Express Future Time   (Chart 3-9)

Directions: Circle all the possible completions.

Example: We   watch / **(are watching) / (are going to watch)**   a news program after dinner.

A. Carmen   **goes / is going / is going to go**   to Mexico City next week. She
            1

   visits / is visiting / is going to visit   her grandparents. Her plane
            2

   leaves / is leaving / is going to leave   at 8:00 A.M. Saturday. She
            3

   stays / is staying / is going to stay   with her grandparents for two weeks.
            4

B. A: What time does Simone's birthday party start tomorrow night?

   B: The party   **starts / is starting / is going to start**   at 7:30.
                        1

   A: I   **bring, am bringing / am going to bring**   a chocolate cake.
                2

   B: She'll love that!

C. The department store   **has / is having / is going to have**   a big sale tomorrow.
                                1

   In fact, the store   **opens / is opening / is going to open**   at 6:00 A.M. Many shoppers
                            2

   go / are going / are going to go   shopping early tomorrow morning. The store
            3

   closes / is closing / is going to close   at 10:00 tomorrow night.
            4

QUIZ 14 Expressing Future Time (Charts 3-1 → 3-9)

Directions: Circle the correct sentence in each pair. The first one is done for you.

1. a. I'm traveling this summer.
   b. I travel this summer.
2. a. Beth retired next June
   b. Beth will retire next June.
3. a. Ruth isn't going to work on Friday.
   b. Ruth not working on Friday.
4. a. Dinner will be ready in ten minutes.
   b. Dinner is being ready in ten minutes.
5. a. It snows tomorrow.
   b. It is going to snow tomorrow.
6. a. The new manager is arriving in a few minutes.
   b. The new manager arrived in a few minutes.
7. a. Mary and Peter get married next year.
   b. Mary and Peter are going to get married next year.
8. a. School starts on September 1st.
   b. School starting on September 1st.
9. a. The kids will be asleep by 8:30 P.M.
   b. The kids will going to sleep by 8:30 P.M.
10. a. The mechanic repairs my car this afternoon.
    b. The mechanic is going to repair my car this afternoon.
11. a. My little brother's clapping his hands when he sees the ice cream truck.
    b. My little brother's going to clap his hands when he sees the ice cream truck.

QUIZ 15 Chapter Review

Directions: Circle the correct completions. The first one is done for you.

SITUATION: *A Week in the Mountains*

My vacation this year **(is)/ will** going to be wonderful. I **am / am going to** about
　　　　　　　　　　　　　　　1　　　　　　　　　　　　　　　　　　　　　2
to leave for a week in the mountains. I **spend / am spending** several days hiking and
　　　　　　　　　　　　　　　　　　　　　　　3
climbing. At night, when I **get / will get** tired, I **find / will find** a place to set
　　　　　　　　　　　　　　　　　　4　　　　　　　　　　5
up my tent. I will build a campfire and **cook / going to cook** my food. Then I
　　　　　　　　　　　　　　　　　　　　　　　　6
am looking / am going to look at the stars through my small telescope.
　　　　　　7
I **maybe / may be** really tired, so I will go to bed early. I **will / am going** sleep very well be-
　　　8　　　　　　　　　　　　　　　　　　　　　　　　　　　　　　9
cause of the fresh air. Then I will feel much better when I **wake / will wake** up in the morning.
　　　　　　　　　　　　　　　　　　　　　　　　　　　　　　　10
I will **probably / may** have a lot of fun on this vacation. I'm really looking forward to it!
　　　　　　11

QUIZ 16 Chapter Review

Directions: Correct the errors.

Example: Tomorrow, I ~~paint~~ *am going to paint* the kitchen.

1. After I am going to feed the children, I will start dinner for the rest of us.

2. In two years, I will quit my job and sailing around the world.

3. You're getting a ticket if you continue to drive so fast.

4. Tina washes the windows this afternoon.

5. Ms. Reed maybe arrive by 10:00 tomorrow.

6. I am go to a conference on early childhood learning next week.

7. Shhh. The baby about to go to sleep.

8. When we going to leave?

9. If I will have time, I will help you.

10. Charlie and Kate is getting married next summer.

QUIZ 17 Chapters 1 → 3 Review

Directions: Complete the sentences with the correct form of the verb in parentheses. Read carefully for time expressions. More than one answer may be correct.

Examples: When the pizza (*come*) ___comes___, let's eat. I (*get*) ___am getting___ hungry.

1. As soon as the weather (*clear*) _____, the plane (*take off*) _____. I hope we (*have*) _____ a smooth flight.

2. Last week, our teacher (*give*) _____ us a surprise quiz. We (*be*) _____ very unhappy.

3. I have a busy day tomorrow. First, I (*take*) _____ my son to get a haircut. Then we (*meet*) _____ friends for a quick lunch. As soon as we (*finish*) _____ lunch, we (*go*) _____ home to bake cookies.

4. A: Who's at the door?
   B: I don't know. I (*check*) _____. There's a package for you.
   A: OK. I (*come*) _____ right now.

5. A: (*you, watch*) _____ the soccer match with us later?
   B: No, thanks. I (*work*) _____ on my essay for history class. I have to turn it in tomorrow.

6. A: Ms. Martin, can we go outside?
   B: No, Johnny. You can't leave until the bell (*ring*) _____.
   A: When (*the bell, ring*) _____?
   B: In five minutes.

7. Tomorrow after work, I (*pick up*) _____ my daughter at her dance class and (*take*) _____ her home. Then we (*eat*) _____ dinner. After dinner, my daughter (*spend*) _____ a few minutes chatting online with her friends before she (*do*) _____ her homework.

CHAPTER 3 – TEST 1

Part A *Directions:* Complete the conversations with **will** and the words in parentheses.

1. A: I don't understand these directions.

   B: I (*explain*) _____ them.

2. A: I'm depressed. I'm not learning English quickly enough.

   B: Don't worry. You (*see, probably*) _____ a lot of improvement soon.

3. A: Don't forget to pick up the clothes at the cleaners.

   B: No problem. I (*forget, not*) _____.

4. A: Does anyone want to go to the store with me?

   B: Rachel (*go*) _____ with you. (*you, get*) _____ some milk?

   A: Sure.

Part B *Directions:* Complete the conversation with **be going to** and the words in parentheses.

5. A: What (*you, do*) _____ on your summer break?

   B: I (*visit*) _____ some friends in Oregon. They live near the ocean.

   A: (*you, stay*) _____ there the entire time?

   B: No, I (*be, probably*) _____ there a few weeks. After that, (*look*) _____ for a part-time job. I want to earn some money for college.

Part C *Directions:* Complete the conversations with **will** or **be going to** and the verbs in parentheses. If there is a plan, use **be going to**.

6. A: I can't find my keys, and I'm in a hurry.

   B: I (*help*) _____ you look for them.

7. A: Do you plan to buy your lunch at school?

   B: No, I (*make*) _____ it at home. It's much less expensive.

8. A: Why are you looking at wedding rings?

   B: I (*ask*) _____ Julie to marry me.

   A: What (*she, say*) _____ ?

   B: I hope (*she, say*) _____ yes!

Part D *Directions:* Circle all the possible completions.

SITUATION: *Love of Soccer*

1. Anne loves soccer! She **is going to play / plays / is playing** the first game of the season next Saturday.

2. The game **starts / will start / is going to start** at 2:30. She's looking forward to it!

3. In a couple of weeks, the team **travels / is traveling / is going to travel** by bus to a soccer tournament.

4. They **drive / are driving / are going to drive** five hours to get there.

5. I think her team **is winning / is going to win / will win** a lot of games this year. All the players on the team are really good!

Part E *Directions:* Complete the conversations with the correct form of the words in parentheses. Use present, past, or future.

1. A: Oh no!

   B: What's the matter?

   A: My computer just *(crash)* _____.

   B: *(you, lose)* _____ all your work?

   A: I don't know. As soon as I *(make)* _____ some repairs, I *(check)* _____.

2. A: What time *(your flight, arrive)* _____ tomorrow?

   B: I'm not sure.

Part F *Directions:* Check (✓) the correct sentences. Correct the incorrect sentences.

1. _____ After we will get married, we will buy a house.

2. _____ The train leaves at 6:45 tomorrow morning.

3. _____ Mari will sing in the choir and going to play the piano at her school concert next week.

4. _____ I go downtown tomorrow with my friends.

5. _____ My husband and I are will not use our credit card so much next month.

6. _____ If Eric call me, I am going tell him I am not available to work this weekend.

7. _____ School ends next week.

8. _____ Tomorrow when John is going to get home, he will help you plant your vegetable garden.

9. _____ Toshi maybe will quit his job soon.

10. _____ Yoko is crying when she hears my news.

CHAPTER 3 – TEST 2

Part A *Directions:* Complete the conversation with *will* and the words in parentheses.

1. A: What's the weather forecast for the rest of the week?

   B: Tomorrow it (*rain*) _____ all day, but after that it (*be*) _____ dry. We (*have, probably*) _____ sun for several days. I believe it (*rain, not*) _____ again until next week.

   A: Great! Let's go to the beach later this week. I (*bring*) _____ a picnic lunch.

   B: That's a good idea!

Part B *Directions:* Complete the conversation with *be going to* and the words in parentheses.

2. A: What (*you, do*) _____ with those beautiful flowers in your garden?

   B: I (*take*) _____ some of them to work and (*put*) _____ them on my desk. I (*give, probably*) _____ the rest to my co-workers and neighbors.

   A: I'm sure they (*like*) _____ that.

Part C *Directions:* Complete the conversations with *will* or *be going to* and the verbs in parentheses. If there is a plan, use *be going to*.

3. A: Do you want to wash or dry the dishes?

   B: I (*dry*) _____ them.

4. A: Why are you wearing a bathing suit?

   B: I (*go*) _____ swimming.

   A: No! It's too cold out.

   B: Don't worry. I (*swim*) _____ indoors, not outdoors!

5. A: Mmmm. Those cookies sure smell good.

   B: Don't touch! We (*give*) _____ them to our new neighbors.

6. A: That was a delicious dinner.

   B: How about some tea?

   A: Sure. But sit down. I (*get*) _____ it.

Part D *Directions:* Circle all the possible completions.

SITUATION: *Our Vegetable Garden*

1. Next weekend my husband and I **are planting / are going to plant / plant** our garden.

2. First, we **are going to pull / are pulling / pull** out all of the weeds.

3. After that, we **will dig up / are going to dig up / digs up** the soil, and then we **are deciding / are going to decide / will decide** what to plant.

4. My husband **will be / is going to be / is being** upset when he sees all the work he has to do.

Part E *Directions:* Complete the conversation with the correct form of the words in parentheses. Use present, past, or future.

A: Where (*you, be*) _____ last night?

B: Sorry. I (*be*) _____ at the office. I (*need*) _____ to finish a project this week.

A: (*you, work*) _____ late again tonight?

B: No. I plan to leave by 6:00. If I need to work late, I (*call*) _____ you.

Part F *Directions:* Check (✓) the correct sentences. Correct the incorrect sentences.

1. ____ Michelle starts her vacation tomorrow afternoon.

2. ____ The business office will closing for one week next month.

3. ____ Dinner is almost ready. The oven timer about go off.

4. ____ Fortunately, our teacher not gonna give us a quiz tomorrow.

5. ____ Next Saturday, Boris will stay home and cleaning out his garage.

6. ____ I have to hurry. We leave for our trip in an hour.

7. ____ Masako buy a new truck next week.

8. ____ Our electric bill is going maybe to increase next month.

9. ____ The plane is going to arrive an hour late. There were delays at the airport.

10. ____ Tomorrow when Pierre will get to work, he is going to interview several candidates for the assistant manager position.

Future Time

CHAPTER 4 Present Perfect and Past Perfect

QUIZ 1 The Past Participle (Chart 4-1)

Directions: Write the past participle form of the regular and irregular verbs. The first one is done for you.

Simple Form	Simple Past	Past Participle
1. want	wanted	*wanted*
2. be	was/were	
3. smile	smiled	
4. go	went	
5. study	studied	
6. come	came	
7. play	played	
8. take	took	
9. tie	tied	
10. stop	stopped	
11. make	made	
12. do	did	
13. find	found	
14. know	knew	
15. have	had	
16. go	went	

QUIZ 2 Present Perfect with Unspecified Time: *Ever* and *Never* (Charts 4-1 and 4-2)

Example: A: (*you, eat, ever*) <u>Have you ever eaten</u> lunch at Rose's Cafe?

B: Yes, we <u>   have   </u>. We (*eat*) <u>   have eaten   </u> lunch there several times.

1. A: (*Cara and Jenn, take, ever*) _____ a tour of the White House?

   B: No, they _____. They (*do, never*) _____ that.

2. A: (*you, go, ever*) _____ to Hawaii?

   B: Yes, I _____. I (*be*) _____ to Maui three times.

3. A: (*Adam, play, ever*) _____ in an orchestra?

   B: No, he _____. He (*play, never*) _____ in an orchestra.

4. A: (*your teacher, give, ever*) _____ a surprise quiz?

   B: Yes, she _____. We (*have*) _____ two surprise quizzes this semester.

5. A: (*Natalia, make, ever*) _____ crêpes flambées?

   B: No, she _____. She (*cook, never*) _____ French food.

QUIZ 3 Present Perfect with Unspecified Time: *Already* and *Yet* (Charts 4-1 and 4-3)

Directions: Look at Miriam's calendar for Tuesday, December 15th. Write present perfect questions using the given words. Then write answers to the questions. Make complete sentences with *yet* and *already*. The first one is done for you.

Tuesday, December 15th	
8:30 A.M.	take kids to school
9:00 A.M.	go to the gym
11:00 A.M.	meet electrician
2:00 P.M.	haircut appointment
3:30 P.M.	pick up kids at school
6:00 P.M.	dinner with the Costas

It is 1:30 P.M. right now.

1. Miriam \ take \ her kids to school \ already?

    Has Miriam already taken her kids to school?

    Yes, she has already taken her kids to school.

2. Miriam \ pick up \ her kids at school \ yet?

3. Miriam \ go \ to the gym \ yet?

4. Miriam \ have \ dinner with the Costas \ already?

5. Miriam \ meet \ with the electrician \ yet?

6. Miriam \ get \ a haircut \ yet?

QUIZ 4 Present Perfect with *Since* and *For* (Charts 4-1 and 4-4)

Directions: Complete each sentence with the present perfect form of the given verb. The first one is done for you.

SITUATION: *Alicia is a travel writer.*

For almost seven years, Alicia ...

1. live   _____*has lived*_____ in New York City.
2. be   _____ a professional writer.
3. work   _____ at *Travel World* magazine.
4. meet   _____ people from every continent through her work.
5. love   _____ her work.

Since she began her job at *Travel World*, Alicia ...

6. visit   _____ the Taj Mahal.
7. go   _____ hiking in the Swiss Alps.
8. eat   _____ sushi in Japan.
9. drink   _____ café au lait in France.
10. see   _____ the Great Wall of China.
11. travel   _____ by boat through the Amazon.

QUIZ 5 Present Perfect with *Since* and *For* (Charts 4-1 and 4-4)

Directions: Complete each sentence with the present perfect form of the verbs in parentheses. The first one is done for you.

SITUATION: *Selena's Favorite Books*

Selena is 15 years old, and she likes to read. She (read) __has read__₁ many books since she was a little girl. She (enjoy) _____₂ both novels and true-life books. For the last several years, her favorites (be) _____₃ the books in the *Traveling Pants* series. Since she read the first book, she (become) _____₄ a big fan of these stories about teen girls and some special jeans. She (visit) _____₅ the series website many times, and she (write) _____₆ email to the books' author. Two years ago, she started decorating her own jeans like the girls in the stories did. For the last two years, she (add) _____₇ a new decoration every time she (have) _____₈ a new and interesting experience. Since she started decorating the jeans, she (draw) _____₉ or (paint) _____₁₀ a rainbow, a music note, a soccer ball, and a heart on them. She (find) _____₁₁ pleasure in this project since she started it. She reads other books, too, but she still enjoys these stories about the girls and their unusual jeans.

QUIZ 6 *Since* vs. *For* (Chart 4-4)

Directions: Complete the sentences with *since* or *for*.

Examples: Kathleen has lived in Dublin …

__since__ 2010.

__for__ her entire life.

Macey has had a part-time job …

1. _____ almost two months.
2. _____ April.
3. _____ a few weeks.
4. _____ last week.
5. _____ several days.
6. _____ the beginning of the year.
7. _____ school started.
8. _____ she moved to the city.
9. _____ a long time.
10. _____ about one year.

QUIZ 7 Since vs. For (Chart 4-4)

Directions: Complete the sentences with *since* or *for*.

SITUATION: *People in My Neighborhood*

Example: Pam has played the violin in the school orchestra __for__ two years.

1. Mr. and Mrs. Nelson have lived in a retirement home _____ June. They have been there _____ five months.

2. Carmen has been a nurse _____ 15 years. She has had a job at the city hospital _____ a few weeks.

3. Leo hasn't worked _____ 2017. He has been out of work _____ he hurt his hands in an accident.

4. Sharon has been in bed _____ three days. She hasn't felt well _____ she ate seafood last weekend.

5. Nathan has been a chef at a hotel _____ he graduated from cooking school. He took a summer off and worked on a fishing boat _____ a few months. He hasn't worked on a boat _____ then.

QUIZ 8 Time Clauses with Since (Chart 4-4)

Directions: Circle the correct completions.

Example: The weather is / (has been) very cold since it snowed last week.

1. I have known how to read since I **was / have been** four years old.
2. Toshi has loved baseball since his father **took / has taken** him to his first game.
3. Ted and Joanne **were / have been** together since they met in high school.
4. Dr. Maas has wanted to work abroad since she **received / has received** her medical degree.
5. The baby **had / has had** a fever since she woke up this morning.
6. Our families have been friends since they **met / have met** on a vacation in the Bahamas.
7. Luke has wanted to be a writer since he **took / has taken** a creative writing class.
8. My car hasn't run well since I **drove / have driven** it in the desert.
9. Marty **rode / has ridden** his bike to work every day since he started his new job.
10. I haven't seen my boss since he **came / has come** to my office last Tuesday.

QUIZ 9 Simple Past vs. Present Perfect (Chart 4-5)

Directions: Circle the correct completions.

Example: My parents ____ married 23 years ago.

    (a.) got   b. have gotten

1. Carol ____ two dogs and a cat when she was a child.

    a. had   b. has had

2. Brian ____ books on the internet for 20 years. He still enjoys it.

    a. sold   b. has sold

3. I ____ zip-lining in the Amazon rainforest twice and plan to go again next year.

    a. went   b. have gone

4. Khalid ____ the keys to his car yet. He's still looking.

    a. didn't find   b. hasn't found

5. I ____ a toothache since this morning.

    a. has had   b. have had

6. Leo ____ any fresh bread at the supermarket. It was sold out.

    a. didn't find   b. hasn't found

7. Ellen ____ afraid of spiders since one fell from the ceiling into her hair.

    a. was   b. has been

8. Mr. Perez ____ golf until he had a stroke at the age of 85.

    a. played   b. has played

9. We ____ Jack since he moved to Toronto.

    a. didn't see   b. haven't seen

10. I didn't know you lived here! How long ____ here?

    a. did you live   b. have you lived

QUIZ 10 Simple Past vs. Present Perfect (Chart 4-5)

Directions: Complete the sentences with the simple past or present perfect form of the verbs in parentheses.

SITUATION: *Food and Drink*

Example: I really like the food at Bamboo Thai restaurant! I (*eat*) ___have eaten___ there three times this month.

1. A: I (*eat, not*) _____ breakfast yet. I (*be, not*) _____ hungry when I got up this morning.

   B: Really? I (*have*) _____ eggs, toast, and three cups of coffee already.

2. A: (*Oscar, be, ever*) _____ to the Mexican restaurant on Aurora Avenue?

   B: Yes, he _____. He (*go*) _____ there for a business lunch last Thursday.

3. Two months ago, we (*try*) _____ some delicious Chinese dumplings called zongzi. We (*have, not*) _____ any dumplings since then.

4. Last semester, Katarina (*win*) _____ a prize for being the top student in her cooking class. In fact, she (*win*) _____ it several times since she started culinary school.

QUIZ 11 Present Perfect Progressive (Chart 4-6)

A. Directions: Complete the conversations with the present perfect progressive form of the verbs in parentheses.

Example: A: Where's Linda?

   B: She's in her bedroom. She (*talk*) ___has been talking___ on the phone since she woke up.

1. A: What's the matter?

   B: Your dog (*bark*) _____ since you left this morning. You need to do something.

2. A: I hear a noise.

   B: It's my cell phone. It (*beep*) _____ all morning because the battery is almost dead.

3. A: I'm so bored.

   B: I know. Professor Adams (*speak*) _____ on the same subject for more than an hour. Will he ever stop?

(continued on next page)

4. A: I'm getting hungry.

   B: Let's stop for lunch. We (drive) _____ for almost two hours. We need a break.

5. A: What's that?

   B: It's my new wireless speaker. I (try) _____ to connect it to my phone for the last 15 minutes. I hope it will sound good!

B. Directions: Make complete sentences. Use the given words and the present perfect progressive.

Example: the baby \ cry \ for an hour   <u>*The baby has been crying for an hour.*</u>

1. how long \ you \ stand \ here?

   _____

2. I \ work \ since 10:00 A.M.

   _____

3. it \ snow \ for two days

   _____

4. how long \ they \ study \ for the final exam?

   _____

5. the taxi driver \ wait \ for ten minutes

   _____

QUIZ 12 Present Progressive vs. Present Perfect Progressive (Chart 4-6)

Directions: Write the present progressive and present perfect progressive forms of the verbs in *italics*. The first one is done for you.

Present Progressive	Present Perfect	Progressive
1. He *rides* the bus.	is riding	has been riding
2. I *read* my email.		
3. You *don't work* hard.		
4. Isabella *teaches* Italian.		
5. They *dance* the samba.		
6. Josh *doesn't study* history.		
7. We *practice* karate.		
8. Students *do* homework.		
9. She *talks* on the phone.		
10. I *don't eat* breakfast.		
11. My friends and I *laugh* a lot.		

QUIZ 13 Present Progressive vs. Present Perfect Progressive (Chart 4-6)

Directions: Complete the conversations. Use the present progressive or present perfect progressive form of the verbs in parentheses.

SITUATION: *Everyone Is Busy*

Example: A: Do you watch *Crime Fighters*? I (*watch*) __have been watching__ it online since the program first started.

B: I (*see, not*) __haven't seen__ it yet. Is it good?

1. A: It's almost time for dinner. Where's Kate?

   B: She (*do*) _____ her homework at the library. She (*study*) _____ at the library since noon.

2. A: Hi! What (*do, you*) _____ ?

   B: I (*watch*) _____ a really good movie. I (*watch*) _____ for ten minutes. Do you want to join me?

(continued on next page)

3. A: Where (go, you) _____?

   B: I'm on my way to the shopping center. Do you want to come?

   A: No, thanks. I (shop) _____ a lot lately.

4. A: Who is on the phone? You (talk) _____ for almost an hour.

   B: I (talk) _____ to my sister. She's going to take a trip to Jamaica!

5. A: Hi! I haven't seen you at work recently.

   B: I know. I (work, not) _____ much since I hurt my foot.

QUIZ 14 Present Perfect Progressive vs. Present Perfect (Chart 4-7)

Directions: Complete the sentences with the present perfect or present perfect progressive form of the verbs in parentheses.

Example: A: Are you going anywhere during the semester break?

   B: Maybe, but I (decide) __haven't decided__ yet.

1. Peter (go) _____ to the dentist several times this month. He's having problems with his teeth.

2. I need to use a different shampoo. I (comb) _____ my hair for 15 minutes, and it's still a mess.

3. A: How is your mother? I (talk, not) _____ to her since we went to a concert together last month.

   B: She is fine, thank you.

4. Look at all the food on the table. It looks like you (cook) _____ all day.

5. Mark is afraid of flying. He (fly, never) _____ on an airplane.

6. A: Did you hear that Alex is quitting his job?

   B: Yes. I (know) _____ about it for a few weeks.

7. Can't you sit down and rest for a minute? You (work) _____ nonstop.

8. A: I'm sorry I'm late for work. I overslept.

   B: I (hear) _____ that excuse from you too many times. You need to be more responsible.

9. Mr. Hale (*work*) _____ in his garden for hours. It looks beautiful!

10. A: Our neighbor has to move out. He (*pay, not*) _____ the rent for six months.

    B: Oh, that's too bad.

QUIZ 15 Past Perfect (Chart 4-8)

Directions: Identify the order of the actions. Write "1" for the first action and "2" for the second action.

Example: Josh turned in his research paper when he had finished writing it.

   a. _2_ Josh turned in his research paper.

   b. _1_ Josh finished writing his research paper.

1. Pat quickly threw his fishing line into the water. A fish had jumped 20 feet from shore.

   a. ____ Pat threw his fishing line into the water.

   b. ____ A fished jumped.

2. After Bill had recovered from his surgery, he felt better than ever.

   a. ____ Bill recovered from his surgery.

   b. ____ He felt better than ever.

3. I offered Cathy a cup of coffee, but she didn't want any. She had already drunk two cups of coffee.

   a. ____ I offered Cathy a cup of coffee.

   b. ____ Cathy drank two cups of coffee.

4. The soccer players hugged each other and cheered their coach. Their teammate had scored the winning goal.

   a. ____ The teammate scored the winning goal.

   b. ____ The players hugged each other.

5. I tried to give Mary my address on the phone, but she had already disconnected.

   a. ____ I tried to give Mary my address.

   b. ____ Mary disconnected.

QUIZ 16 Past Perfect (Chart 4-8)

Directions: Complete the sentences. Use the past perfect form of the verbs in parentheses.

Example: As soon as I got home from France, I gave Julia her gift. I (*promise*) _____had promised_____ to bring her some perfume from Paris.

1. I finished work late. By the time I got to my bus stop, the bus (*leave, already*) _____.

2. We (*think, not*) _____ of moving until we found out that the schools in the other part of town were much better than ours.

3. Chris and Diane asked me to join them for dinner, but I (*eat, already*) _____.

4. We wanted to build a campfire when we got to our campsite, but the camp ranger (*put up, already*) _____ a sign saying "No Fires."

5. Pedro planned to get an autograph from one of the baseball players, but by the time he arrived, the players (*start*) _____ the pre-game practice.

6. My parents were interested in buying the Clarks' house on Fifth Street, but when they went to look at it, it (*sell, already*) _____.

7. Ken was about to pay his credit card bill when he realized that his wife (*pay, already*) _____ it.

8. Jan was going to talk to her assistant about the importance of arriving at work on time, but her manager (*meet, already*) _____ with him.

9. Hannah sat down on the sofa and began to read the mystery novel. After a few pages, she realized she (*read, already*) _____ it.

10. This month's electric bill was high because we (*leave*) _____ some lights on in the house while we were away on vacation.

QUIZ 17 Chapter Review

Directions: Correct the errors.

SITUATION: *Rita's Trips to the Philippines*

Example: Rita lives in Seattle. She ~~had~~ *has* lived in the United States for 20 years.

1. Rita has left two weeks ago to visit her family in the Philippines.

2. Rita's parents live in Manila. She hadn't been visiting them for two years.

3. Her sister and brother-in-law have been lived in Manila for many years, too.

4. Rita likes to visit Rizal Park when she is in Manila. She had been there many times already.

5. Rizal Park had some beautiful Japanese and Chinese gardens.

6. Every time she has gone to Manila, she has been taking her niece to the Star City Amusement Park.

7. A new aquarium, Manila Ocean Park, opened recently. Rita haven't visited it yet.

8. Since they updated the Manila Baywalk several years ago, it had become Rita's favorite place in Manila.

9. The Baywalk has many outdoor restaurants, cafés, and bars. Rita has been gone there many times in the evenings.

10. Rita's parents had come to the U.S. several times, but Rita always enjoys going "home" to the Philippines.

QUIZ 18 Chapters 1 → 4 Review

Directions: Circle the correct completions.

Example: I can't come with you. I need to stay here. I ____ for a phone call.

    a. wait   b. will wait   (c.) am waiting   d. have waited

1. I ____ my glasses three times so far this year. One time I dropped them on a cement floor. Another time I sat on them. This time I stepped on them.

    a. broke   b. was breaking   c. have broken   d. have been breaking

2. Kate reached to the floor and picked up her glasses. They were broken. She ____ on them.

    a. stepped   b. had stepped   c. was stepping   d. has stepped

3. Sarah gets angry easily. She ____ a bad temper ever since she was a child.

    a. has   b. will have   c. had   d. has had

4. Now, whenever Sarah starts to lose her temper, she ____ a deep breath and ____ to ten.

    a. takes … counts   c. took … counted

    b. has taken … counted   d. is taking … counting

5. Nicky, please don't interrupt me. I ____ to Grandma on the phone. Go play with your trucks so we can finish our conversation.

    a. talk   b. have talked   c. am talking   d. have been talking

6. We ____ at a hotel in Miami when the hurricane hit southern Florida last month. As soon as the hurricane moved out of the area, we left and went back home.

    a. had stayed   b. stay   c. were staying   d. stayed

7. Now listen carefully. When Aunt Martha ____ tomorrow, give her a big hug.

    a. arrives   b. will arrive   c. arrived   d. is going to arrive

8. My cousin ____ with me in my apartment for the last two weeks. I'm ready for him to leave, but he seems to want to stay forever. Maybe I should ask him to leave.

    a. is staying   b. stayed   c. was staying   d. has been staying

9. Mrs. Larsen discovered a bird in her apartment. It was in her living room. It ____ into her apartment through an open window.

    a. was flying   b. had flown   c. has flown   d. was flown

(continued on next page)

10. The phone rang, so I ____ it up and ____ hello.
    a. picked ... had said
    b. picked ... said
    c. was picking ... said
    d. was picking ... had said

11. My mother began to drive when she was 14. Now she is 89, and she still drives. She ____ for 75 years.
    a. was driving   b. drives   c. drove   d. has been driving

12. Since prehistoric times, people in every culture ____ jewelry.
    a. wear   b. wore   c. have worn   d. had worn

13. It ____ when I left the house this morning, so I wore my hat and mittens.
    a. snowed   b. snows   c. is snowing   d. was snowing

14. Australian koala bears are interesting animals. They ____ practically their entire lives in trees without ever coming down to the ground.
    a. are spending
    b. have been spending
    c. have spent
    d. spend

15. The teacher is late today, so class hasn't begun yet. After she ____ here, class will begin.
    a. will get   b. is going to get   c. gets   d. is getting

16. It's raining hard. It ____ an hour ago and ____ yet.
    a. had started ... doesn't stop
    b. has started ... didn't stop
    c. started ... hasn't stopped
    d. was starting ... isn't stopping

17. Alex's bags are almost ready for his trip. He ____ for Austria later this afternoon. We'll say good-bye to him before he ____.
    a. left ... went
    b. leaves ... will go
    c. is leaving ... goes
    d. has left ... will go

18. I heard a slight noise, so I walked to the front door to investigate. I looked down at the floor and saw a piece of paper. Someone ____ a note under the door to my apartment.
    a. had pushed   b. is pushing   c. has pushed   d. pushed

19. I walked slowly through the market. People ____ all kinds of fruits and vegetables. I studied the prices carefully before I decided what to buy.
    a. have sold   b. sell   c. had sold   d. were selling

20. I really like my car. I ____ it for six years. It runs beautifully.
    a. have   b. have had   c. had   d. have been having

CHAPTER 4 – TEST 1

Part A *Directions:* Write the past participle form of the verbs.

1. pay _____
2. swim _____
3. know _____
4. wait _____
5. study _____

6. tell _____
7. leave _____
8. fly _____
9. begin _____
10. eat _____

Part B *Directions:* Complete the sentences with *since* or *for*.

Rosa has worked as a project manager …

1. _____ two years.
2. _____ January.
3. _____ Monday.
4. _____ a few months.
5. _____ last week.
6. _____ a long time.

Part C *Directions:* Complete the sentences. Use the simple past or present perfect form of the verbs in parentheses.

SITUATION: *A New Town*

1. We (*move*) _____ here a year ago. Since then I (*meet*) _____ many nice people.

2. I go to Franklin High School. Since I started high school, I (*have*) _____ some good teachers and some terrible teachers. Last year my favorite teacher (*be*) _____ Mr. Freeman, my chemistry teacher.

3. I'm in our school band. I (*play*) _____ the drums in the last four school concerts. I love it!

Part D *Directions:* Complete the conversations with the words in parentheses. Use the present perfect or the present perfect progressive form.

1. A: (*you, try, ever*) _____ to make homemade chocolate?

   B: Homemade chocolate? No, I _____, but I

   (*make*) _____ chocolate cookies several times.

2. A: (*you, finish*) _____ your homework yet?

   B: Yes, we _____. We just finished.

3. A: The dog (*scratch*) _____ at the door for five minutes.

   B: I think she wants to go outside.

4. A: (*Anna, drive*) _____ the car alone yet?

   B: No, she _____. She's still learning to drive.

5. A: Your eyes are red. (*you, cry*) _____?

   B: No, I have allergies.

Part E *Directions:* Circle the correction completions.

1. Look! The storm ____ branches from the trees onto our lawn. It will take a long time to clean up.

   a. has blown   b. had blown

2. Susie was upset. Another child ____ her toy train. She found it on the floor in several pieces.

   a. has broken   b. had broken

3. I ____ really tired all week. I've been getting enough sleep, so maybe I should see a doctor.

   a. have felt   b. had felt

4. Dina didn't need a ride home. She ____ to work.

   a. has driven   b. had driven

5. When Mark arrived at the station, the train wasn't there. It ____ already.

   a. has left   b. had left

Part F *Directions:* Correct the errors.

1. Gary is at work since 5:00 this morning.

2. Steve have enjoyed listening to Mozart since he took a music class in high school.

3. Nadia hadn't finished her dinner yet. She can't have dessert until she does.

4. It has been raining on my birthday every year for the last ten years.

5. Already I have decided to major in marine biology.

CHAPTER 4 – TEST 2

Part A *Directions:* Write the past participle form of the verbs.

1. visit _____
2. speak _____
3. think _____
4. give _____
5. shop _____

6. buy _____
7. come _____
8. read _____
9. teach _____
10. find _____

Part B *Directions:* Complete the sentences with *since* or *for*.

Mariko has been trying to lose weight …

1. _____ a year.
2. _____ several months.
3. _____ she had a baby.
4. _____ April.
5. _____ a long time.
6. _____ last summer.

Part C *Directions:* Complete the sentences. Use the simple past or present perfect form of the verbs in parentheses.

SITUATION: *Retirement*

1. Greg and Sharon Smith were teachers for many years. Five years ago, they (*quit*) _____ their jobs. They (*travel*) _____ to many countries since they retired.

2. The Smiths (*be*) _____ to Asia several times. Last year, they (*visit*) _____ Vietnam and Thailand.

3. During the past five years, they (*enjoy*) _____ seeing so many different parts of the world.

Part D *Directions:* Complete the conversations. Use the present perfect or present perfect progressive form of the verb in parentheses.

1. A: (*you, wear, ever*) _____ a tuxedo?

   B: No, I _____. They don't look very comfortable.

2. A: Billy, your clothes are all dirty. (*you, jump*) _____ in mud puddles?

   B: Yes. I'm sorry, Mommy.

3. A: This bee (*fly*) _____ around me for ten minutes.

   B: Why don't we go inside?

4. A: (*you, get, ever*) _____ a speeding ticket?

   B: Yes, I _____. I (*get*) _____ two tickets.

5. A: (*Bill, do*) _____ the wash yet?

   B: Yes, he _____. He just finished.

Part E *Directions:* Circle the correct completions.

1. I woke up tired, but I didn't take a nap because I ____ for ten hours.

   a. have already slept   b. had already slept

2. Jason ____ in the ocean. He's going to try it for the first time tomorrow.

   a. has never swum   b. had never swum

3. Dennis ____ his ankle again. He needs to be more careful when he plays soccer.

   a. has hurt   b. had hurt

4. I wanted to buy some new dishes yesterday, but I decided I ____ too much money on clothes.

   a. have already spent   b. had already spent

5. Dr. Brooks takes care of patients, and she ____ two medical devices that help people with heart disease.

   a. has invented   b. had invented

Part F *Directions:* Correct the errors.

1. Sandy have been trying to call you. Did she reach you yet?

2. Ted hadn't called yet. I wonder if he lost our phone number.

3. Andy is on vacation since Saturday.

4. I have been knowing about those problems for a few weeks.

5. Chris has been starting his Ph.D. thesis several times. He's not sure about his topic.

CHAPTER 5 Asking Questions

QUIZ 1 Yes / No Questions and Short Answers   (Chart 5-1)

Directions: Circle the correct verbs.

SITUATION: *Going Fishing*

Example: A: **(Is) / Does**   that your new fishing rod?
            B: Yes, it **(is) / does**.

1. A: **Are / Do**   you like to catch fish?

    B: Yes, I **am / do**.

2. A: **Are / Do**   you going fishing tomorrow?

    B: Yes, I **am / do**.

3. A: **Is / Does**   Pat going to go with you?

    B: Yes, he **is / does**.

4. A: **Were / Did**   you go fishing yesterday?

    B: Yes, we **were / did**.

5. A: **Were / Did**   there many fish in the river?

    B: No, there **weren't / didn't**.

6. A: **Have / Were**   you ever caught a fish before?

    B: No, I **haven't / wasn't**.

7. A: **Was / Did**   Pat catch a fish yesterday?

    B: Yes, he **was / did**.

8. A: **Will / Is**   the weather be nice tomorrow?

    B: Yes, it **will / is**.

9. A: **Is / Are**   there a good place to fish near here?

    B: Yes, there **is / are**.

10. A: **Are / Do**   you and Pat plan to catch lots of fish?

    B: Yes, we **are / do**.

QUIZ 2 Yes / No Questions and Short Answers (Chart 5-1)

Directions: Use the information in parentheses to make *yes/no* questions. Complete each conversation with an appropriate short answer. Do not use a negative verb in the question.

SITUATION: *My Sister's Wedding*

Example: A: __Is your sister going to get married soon?__
B: Yes, __she is__. (My sister is going to get married soon.)

1. A: _____?
   B: Yes, _____. (I like her fiancé very much.)

2. A: _____?
   B: No, _____. (They didn't meet in school.)

3. A: _____?
   B: Yes, _____. (All the wedding plans have been made.)

4. A: _____?
   B: Yes, _____. (They are inviting many people.)

5. A: _____?
   B: No, _____. (Mr. and Mrs. Jennings won't be at the wedding.)

6. A: _____?
   B: Yes, _____. (I will sing at the wedding.)

7. A: _____?
   B: No, _____. (I am not nervous.)

8. A: _____?
   B: Yes, _____. (They are going to move to an apartment.)

9. A: _____?
   B: No, _____. (My sister hasn't bought her dress yet.)

10. A: _____?
    B: Yes, _____. (You asked too many questions.)

QUIZ 3 Where, Why, When, and What Time (Chart 5-2)

Directions: Circle the correct question words to complete the conversations. The answers to the questions are in parentheses.

Example: ___ do you usually go to bed? (*Around midnight.*)

    a. Where   b. Why   (c.) What time

1. ___ does the party start? (*At 9:30.*)

    a. What time   b. Where   c. Why

2. ___ will the party be? (*At Sam's.*)

    a. What time   b. Where   c. Why

3. Professor, ___ are you leaving? (*Because I have a class in ten minutes.*)

    a. when   b. where   c. why

4. ___ can I find out my test results? (*Tomorrow morning.*)

    a. When   b. Where   c. Why

5. ___ is a good place for us to meet tomorrow? (*At the Mountain View Café.*)

    a. When   b. Where   c. Why

6. ___ shall we meet? (*10:00 A.M.*)

    a. What time   b. Where   c. Why

7. ___ did Dr. Smith call? (*About an hour ago.*)

    a. When   b. Where   c. Why

8. ___ are you going to the hospital? (*On Friday.*)

    a. What time   b. Where   c. When

9. ___ did Yasuko move? (*Because her rent was too high.*)

    a. What time   b. Where   c. Why

10. ___ does she live now? (*On Fourteenth Avenue West.*)

    a. Where   b. When   c. Why

QUIZ 4 Where, Why, When, What Time, How Come, and What ... For (Chart 5-2)

Directions: Complete two questions using the information from each sentence.

SITUATION: *A Trip to Greece*

Examples: Sven and Erik are going to the airport at 9:30.

    What time _are Sven and Erik going to the airport?_

    Where _are Sven and Erik going at 9:30?_

1. Sven and Erik are flying to Greece on June 28th.

    Where _____?

    When _____?

2. Their flight will arrive in Athens on June 29th at about 1:00 P.M.

    What time _____?

    When _____?

3. Sven wants to go to Athens because he wants to see the Acropolis.

    Where _____?

    What _____ for?

4. Erik was in Greece five years ago because he was a student there.

    When _____?

    How come _____?

5. Sven and Erik will return home on July 15th because they have to go back to work.

    Where _____?

    Why _____?

Asking Questions 73

QUIZ 5 Questions with Who and What (Chart 5-3)

Directions: Complete the conversations with **who** or **what**.

Example: A: <u>Who</u> is at the door?

B: It's our neighbor.

1. A: _____ did you buy?

   B: We bought a new kitchen table.

2. A: _____ did Neil send the text to?

   B: He sent it to his best friend.

3. A: _____ is that?

   B: It's a honeydew melon.

4. A: _____ did you eat lunch with?

   B: I ate with Alison.

5. A: _____ do the kids wear to school?

   B: They wear blue and white uniforms.

6. A: _____ lives next door to you?

   B: The Johnsons live there.

7. A: _____ was the call from?

   B: It was from my sister.

8. A: _____ does Norita eat for breakfast?

   B: She usually has toast and coffee.

9. A: _____ is your doctor?

   B: His name is Dr. Griggs.

10. A: _____ baked this delicious apple pie?

    B: My husband.

QUIZ 6 Questions with *Who* and *What* (Chart 5-3)

Directions: Make questions with *who* or *what*. Use the same verb tense as the answer in parentheses.

Example: <u>Who painted the picture?</u> Picasso. (*Picasso painted the picture.*)

1. _____

   Henry. (*Bill saw Henry.*)

2. _____

   Roberto. (*Roberto saw the bear.*)

3. _____

   The clerk. (*Marcella paid the clerk.*)

4. _____

   A hamburger. (*Charles ordered a hamburger.*)

5. _____

   Kim. (*Kim rides a motorcycle.*)

6. _____

   Mr. Brown. (*Mr. Brown came late.*)

7. _____

   A dog. (*Ruth brought home a dog.*)

8. _____

   Lee's toy train. (*Lee's toy train broke.*)

9. _____

   Gina. (*Laura called Gina.*)

10. _____

    Juan's parents. (*Juan's parents won a contest.*)

QUIZ 7 Using *What* and a Form of *Do* (Chart 5-4)

Directions: Make questions beginning with *what* and a form of *do* to complete the conversations. Use the information in *italics* to help you.

Example: A: _____*What did you do*_____ yesterday afternoon?

B: We *went* hiking near Mount Shasta yesterday afternoon.

1. A: _____ when your baby cries?

   B: I *sing* to my baby when she cries.

2. A: _____ tomorrow?

   B: We *are going to go to the art museum* tomorrow.

3. A: _____ after dinner?

   B: Anne usually *washes the dishes and puts them away* after dinner.

4. A: _____ when you saw the accident?

   B: I *stopped my car and called 911* when I saw the accident.

5. A: _____ if it snows?

   B: We*'ll stay home* if it snows.

6. A: _____ this afternoon?

   B: I *would like to take a nap* this afternoon.

7. A: _____ when you turn 65?

   B: I *want to retire* when I turn 65.

8. A: _____ right now?

   B: I*'m fixing my bike* right now.

9. A: _____ last weekend?

   B: Jason *went to Las Vegas* with a friend last weekend.

10. A: _____ for a living?

    B: Geologists *study rocks*.

QUIZ 8 Which vs. What (Chart 5-5)

Directions: Complete the conversations with *which* or *what*.

Example: A: Can I please have an ice cream cone?

B: We have vanilla and chocolate. __What__ kind do you want?

A: Vanilla, please.

1. A: There are three different kinds of soup on the menu.

    B: _____ one do you want?

    A: I guess I'll have vegetable soup, please.

2. A: I found a wallet in the street near my house.

    B: _____ are you going to do with it?

    A: I'll probably take it to the police station.

3. A: _____ kind of movies does Raymond like?

    B: He enjoys comedies.

4. A: I know you can write with both your left and right hands, but _____ do you prefer to use?

    B: My left hand.

5. A: Tom has lost a lot of weight!

    B: Yeah, he has. I don't know _____ size he wears now.

6. A: _____ kind of pasta should we have for dinner?

    B: Let's make spaghetti carbonara. It's my favorite.

7. A: I like all of these dresses.

    B: All four dresses look nice on you. I don't know _____ one you should buy.

    A: I think I'll get the black one.

8. A: _____ color eyes do you have? They seem to change color with the clothes you wear.

    B: They are blue, but sometimes they look green.

9. A: There are several websites with local news. _____ one is the best?

    B: I like ShorelineAreaNews.org. It's very balanced.

10. A: _____ score did you get on your driving test?

    B: I don't know exactly, but I passed!

QUIZ 9 Questions with *How* (Chart 5-6)

Directions: Match each question with the best response. The first one is done for you.

1. _f_ How tired are you?
2. ___ How old is the baby?
3. ___ How tall is your father?
4. ___ How big is your apartment?
5. ___ How hot is it outside?
6. ___ How good is that book?
7. ___ How did you do on the test?
8. ___ How soon will this movie end?
9. ___ How hungry are you?
10. ___ How well do you play basketball?
11. ___ How do you get to work?

a. Very interesting.
b. By car.
c. In ten minutes.
d. Just three months.
e. I got 83 percent.
f. I'm about to fall asleep.
g. It has four rooms.
h. I'm starving.
i. Six feet.
j. About 90° F.
k. I am just learning.

QUIZ 10 Questions with *How Often, How Many, How Far,* and *How Long* (Charts 5-7 → 5-9)

Directions: Complete the conversation questions with *often, many, far,* or *long*.

Example: SUZANNE: How __long__ have you lived here?

   LUCIO: About 25 years.

1. ELENA: Sara, how _____ do you see your grandchildren?

   SARA: About once a week.

   ELENA: How _____ away do they live?

   SARA: About 35 miles.

   ELENA: How _____ does it take to get to their house?

   SARA: If traffic isn't heavy, it takes about an hour.

2. GIL: Max, how _____ times a week do you exercise?

   MAX: Three to four times a week. I work out at the gym, and I run.

   GIL: How _____ do you spend at the gym?

   MAX: Usually an hour.

   GIL: How _____ do you run?

   MAX: Five to 10 kilometers, depending on how much time I have.

3. KENTO: Hiro, how _____ miles is it from your apartment to your office?

   HIRO: About ten miles.

   KENTO: How _____ does it take you to get to work by car?

   HIRO: About 45 minutes.

   KENTO: How _____ do you take the train to work?

   HIRO: Almost every day.

   KENTO: And how _____ times a month do you drive?

   HIRO: Only once or twice a month—when I am running late.

QUIZ 11 Review of How (Charts 5-6 → 5-9, 5-11, and 5-12)

Directions: Make complete questions with **how** for the given answers. Use the information in *italics* to help you.

Example: A: <u>How soon do we need to leave?</u>

B: We need to leave **in ten minutes**.

1. A: _____

   B: It took me about **30 minutes** to check my email.

2. A: _____

   B: **E - L - E - P - H - A - N - T**.

3. A: _____

   B: We go to the movies about **three times a month**.

4. A: _____

   B: I came here **by bus**.

5. A: _____

   B: Our school is just **two blocks** away.

6. A: _____

   B: This water is **very cold**. You can't swim in it.

7. A: _____

   B: "Ms." is an easy word **to pronounce**. You say "mizz."

8. A: _____

   B: Jon's birthday was yesterday. He's **20**.

9. A: _____

   B: I'm feeling **great**!

10. A: _____

    B: Mr. Wang speaks English **perfectly**. He sounds like a native speaker.

QUIZ 12 Information Questions Review (Charts 5-2 → 5-12A)

A. Directions: Write information questions. Use *where, what, who, why, how,* or *how long.* Use the information in *italics* to help you. The first one is done for you.

SITUATION: *A Job Interview*

Emily is at a job interview. Write the questions the interviewer asks her.

1. INTERVIEWER: Good morning. _How are you doing today?_

   EMILY: I'm doing *fine*, thank you.

2. INTERVIEWER: _____

   EMILY: I want to work here *because this is a good company*.

3. INTERVIEWER: _____

   EMILY: Now I work *at KRW Enterprises* as a computer programmer.

4. INTERVIEWER: _____

   EMILY: I have worked there *for five years*.

5. INTERVIEWER: _____

   EMILY: My favorite project was *writing a program to help students learn English*.

6. INTERVIEWER: _____

   EMILY: My supervisor is *Joshua Smith*.

B. Directions: Write information questions. Use *how, when, who, how far,* or *how often.* Use the information in *italics* to help you.

SITUATION: *Questions in Class*

A teacher is asking her students questions about their summer vacation. Write the questions she asks.

1. TEACHER: _____

   SUE: *Ian* had the most exciting vacation. He rode a camel in Morocco!

2. TEACHER: _____

   IAN: We rode *for about ten kilometers*.

3. TEACHER: _____

   IAN: *It felt strange* at first, but then it was fun!

4. TEACHER: _____

   JILL: We went to the beach *in July*.

5. TEACHER: _____

   JILL: We go to the beach almost *every summer*.

QUIZ 13 Tag Questions (Chart 5-13)

Directions: Complete the tag questions with the correct verbs.

SITUATION: *A Rainy Day*

Example: The weather is terrible today, _____isn't_____ it?

1. You are going out, _____ you?
2. You're not sick, _____ you?
3. You have a coat on, _____ you?
4. You are warm enough, _____ you?
5. It's not snowing, _____ it?
6. You don't want to walk, _____ you?
7. Pam is going with you, _____ she?
8. Your brother doesn't want to go with you, _____ he?
9. The Petersons are out walking in the rain, _____ they?
10. They don't have an umbrella, _____ they?

QUIZ 14 Tag Questions (Chart 5-13)

Directions: Complete the tag questions with the correct verbs.

SITUATION: *World Travelers*

Example: The Jamisons traveled a lot this year, _____didn't_____ they?

1. They've been around the world twice, _____ they?
2. They haven't been to New Zealand, _____ they?
3. One trip was canceled, _____ it?
4. They had to change their plans, _____ they?
5. They didn't go to Italy, _____ they?
6. They are in China now, _____ they?
7. They sent you a postcard, _____ they?
8. You haven't talked to them recently, _____ you?
9. It's been a good year for them, _____ it?
10. You will see them soon, _____ you?

QUIZ 15 Chapter Review

Directions: Circle the correct question in each pair. The first one is done for you.

1. a. Which kind of cell phone do you like?

   (b.) What kind of cell phone do you like?

2. a. How long does it take to fly from London to Moscow?

   b. What time does it take to fly from London to Moscow?

3. a. I'm hungry. How about you?

   b. Do you hungry?

4. a. When your plane arrives from Paris?

   b. What time does your plane arrive?

5. a. How does it going?

   b. How are you doing?

6. a. How far is it from Los Angeles to San Francisco?

   b. How long is it from Los Angeles to San Francisco?

7. a. Where are you going on vacation?

   b. Where you are going on vacation?

8. a. Does John has time to help me?

   b. Will John have time to help me?

9. a. When will Naomi finish her driving lesson?

   b. What time Naomi finishes her driving lesson?

10. a. It's raining, isn't it?

    b. It's not going to rain, isn't it?

11. a. How come did you do that?

    b. What did you do that for?

CHAPTER 5 – TEST 1

Part A *Directions:* Complete the questions with *how, how far, how long, how often, how soon, what, when, which, who,* or *why.*

1. A: _____ do you like this town?

    B: We like it a lot.

2. A: Traffic was terrible this morning. _____ did it take you to get to work?

    B: About an hour.

3. A: _____ will Pam be here?

    B: She'll be right back. She just went to get a drink of water.

4. A: _____ is picking you up from school this afternoon?

    B: My older brother.

5. A: _____ did it take you to finish the test?

    B: About an hour.

6. A: These are the two bookcases we can afford. _____ one would be better?

    B: The shorter one.

7. A: _____ is Billy crying?

    B: He fell down and cut his knee.

8. A: _____ do you go to the coffee shop?

    B: About twice a week.

9. A: _____ are you doing?

    B: I'm trying to download a new app on my phone.

10. A: _____ is it from here to downtown?

    B: About six kilometers.

Part B *Directions:* Complete the conversations with information questions. Use the information in *italics* to help you.

SITUATION: *Sara Is a Runner*

1. A: _____

   B: Sara trains *an hour a day*.

2. A: _____

   B: She runs *30 kilometers a week*.

3. A: _____

   B: She runs *very* fast.

4. A: _____

   B: She is going to run in a ten-kilometer race *next week*.

5. A: _____

   B: She will run with *several friends*.

6. A: _____

   B: She plans *to win*.

7. A: _____

   B: She spells her name *S - A - R - A*.

Part C *Directions:* Complete the tag questions with the correct verbs.

1. You work outdoors, _____ you?
2. It didn't rain last night, _____ it?
3. Jane's been very busy, _____ she?
4. We aren't going to attend the meeting, _____ we?
5. Sam went to the doctor again, _____ he?
6. Your cat is friendly, _____ it?

Part D *Directions:* Correct the errors.

1. What you know about the new department manager?
2. Why that dog is barking so loudly?
3. What means "besides"?
4. Marta changed jobs last month, wasn't she?
5. Which movie you see last night, *Monsters* or *Dragons*?
6. How long it takes to fly from Paris to London?
7. Who did tell you about the party?

CHAPTER 5 – TEST 2

Part A *Directions:* Complete the questions. Use a word or phrase from the box.

what	who	when	why	which
how	how far	how long	how soon	how often

SITUATION: *Interview with a University Student*

1. A: _____ are you living right now?

   B: In an apartment near campus.

2. A: _____ is it from here to your apartment?

   B: About three miles.

3. A: _____ does it take you to get here?

   B: Twenty minutes.

4. A: _____ did you come here?

   B: Because I wanted to study at a university.

5. A: _____ do you go back home?

   B: About twice a year.

6. A: _____ do you like it here?

   B: I love it!

7. A: _____ is helping you pay for school?

   B: My parents.

8. A: _____ do you plan to graduate?

   B: In two years.

9. A: I've forgotten. _____ major did you decide on, math or business?

   B: Business.

10. A: _____ will your parents visit you?

    B: Next December.

Part B *Directions:* Complete the questions with *how, how far, how long, how soon, how often, what, who, when, which,* or *why*. Use the information in *italics* to help you.

SITUATION: *Jill Has a New Job*

1. A: _____

   B: Jill got a new job *last week*.

2. A: _____

   B: She works *at Acme Nova Software*.

3. A: _____

   B: She is going to *develop marketing plans* for the company.

4. A: _____

   B: She plans to be there for *at least two years*.

5. A: _____

   B: She drives to work *three times a week*.

6. A: _____

   B: She goes to work *by bus* on the other days.

7. A: _____

   B: She works *60 hours* a week.

Part C *Directions:* Complete the tag questions with the correct verbs.

1. You got home late last night, _____ you?

2. It isn't raining right now, _____ it?

3. We don't have to leave yet, _____ we?

4. Dr. Wilson had to leave early, _____ she?

5. Otto is going to pay me back, _____ he?

6. They're not lost, _____ they?

7. You don't like this movie, _____ you?

Part D *Directions:* Correct the errors.

1. What means "anyway"?

2. Sonya needs more time, isn't she?

3. What kind soup you want for lunch?

4. How you say "good morning" in Spanish?

5. Why you were late today?

6. How long does it takes to get from here to your home?

CHAPTER 6 — Nouns and Pronouns

QUIZ 1 Plural Forms of Nouns   (Chart 6-1)

Directions: Write the singular or plural form of the given words. The first one is done for you.

1. one table — many _tables_
2. a mouse — six _____
3. one _____ — a lot of leaves
4. a city — two _____
5. one _____ — several tomatoes
6. a box — some _____
7. a woman — a lot of _____
8. a _____ — two sheep
9. one tooth — several _____
10. one child — many _____
11. a business — two _____

QUIZ 2 Pronunciation of Final -s / -es   (Chart 6-2)

Directions: Circle the sound at the end of each word. The first one is done for you.

1. hands   /s/   (/z/)   /əz/
2. messages   /s/   /z/   /əz/
3. cats   /s/   /z/   /əz/
4. skills   /s/   /z/   /əz/
5. boys   /s/   /z/   /əz/
6. mistakes   /s/   /z/   /əz/
7. boxes   /s/   /z/   /əz/
8. shirts   /s/   /z/   /əz/
9. phones   /s/   /z/   /əz/
10. laptops   /s/   /z/   /əz/
11. dishes   /s/   /z/   /əz/

QUIZ 3 Subjects, Verbs, and Objects   (Chart 6-3)

Directions: Complete each diagram with the correct subject, verb, and object. Write *Ø* if there is no object.

Example: The woman bought some sunglasses.

The woman	*bought*	*some sunglasses*
subject	verb	object of verb

Her eyes were tired.

Her eyes	*were*	*Ø*
subject	verb	object of verb

1. Steve asked a question.

Steve	asked	a question
subject	verb	object of verb

2. His question wasn't clear.

His question	wasn't	Ø
subject	verb	object of verb

3. My phone rang three times.

My phone	rang	Ø
subject	verb	object of verb

4. I answered the phone.

I	answered	the phone
subject	verb	object of verb

5. Hanifa loves animals.

Hanifa	loves	animals
subject	verb	object of verb

6. She has had many pets.

She	has had	many pets
subject	verb	object of verb

7. The police stopped several cars.

The police	stopped	several cars
subject	verb	object of verb

8. The drivers looked surprised.

The drivers	looked	Ø
subject	verb	object of verb

9. My daughter goes to college.

My daughter	goes	Ø
subject	verb	object of verb

10. She is studying anthropology.

She	is studying	anthropology
subject	verb	object of verb

QUIZ 4 Subjects, Verbs, and Objects (Chart 6-3)

Directions: If the word in ***italics*** is used as a noun, circle "N." If it is used as a verb, circle "V."

Examples: My sister usually **cooks** breakfast on Sunday mornings.   N   (V)
There are three good **cooks** in my family.   (N)   V

1. a. Seattle has many lovely ***parks***.   N   V

   b. My neighbor usually ***parks*** his car in the driveway.   N   V

2. a. I didn't ***reply*** to the text.   N   V

   b. I didn't expect a ***reply*** from my friend.   N   V

3. a. My daughter ***plays*** basketball after school every day.   N   V

   b. We saw two very good ***plays*** at the theater.   N   V

4. a. There is a bus ***stop*** about two blocks away.   N   V

   b. Careful drivers always ***stop*** at red lights.   N   V

5. a. The kids ***laugh*** a lot when they watch cartoons.   N   V

   b. Richard has a very loud ***laugh***.   N   V

QUIZ 5 Objects of Prepositions (Charts 6-3 and 6-4)

Directions: Read the sentences. Write "P" over the prepositions and "Obj. of P." over the objects of prepositions.

SITUATION: *Snowy Winter*

Example: The little boy slid quickly down the snowy hill.
(P = down, Obj. of P = hill)

1. Winter days are short. It's dark when I go to school, and it also gets dark early during the winter months.

2. Children love playing outside in the snow. They throw snowballs at each other and make snowmen.

3. Winter sports are popular with many people. They go skiing and snowboarding in the mountains.

4. Winter can sometimes be dangerous. Last winter, two teenagers fell into a frozen lake near their home. Rescuers found them under the ice and pulled them to safety.

QUIZ 6 Prepositions of Time   (Chart 6-5)

Directions: Complete the sentences with *in, at,* or *on*. The first one is done for you.

The earthquake occurred …

1. _at_ midnight.
2. _____ 12:00.
3. _____ the spring.
4. _____ May 24th.
5. _____ September.
6. _____ Wednesday night.
7. _____ the twenty-first century.
8. _____ 2017.
9. _____ a weekend.
10. _____ the early morning.
11. _____ Saturday.

QUIZ 7 Subject and Verb Agreement   (Chart 6-7)

Directions: Circle the correct verbs.

Example: There (is) / are   only one library in my hometown.

1. Every student at the university **lives / live** in the dorm for the first year.
2. The raindrops on the trees **is / are** sparkling in the sunlight.
3. Here **is / are** a used laptop for sale.
4. Do you think that everyone **needs / need** a cell phone?
5. Some spellings in English **doesn't / don't** seem to make sense.
6. People everywhere **uses / use** the internet to get information.
7. The mountains in the distance **is / are** popular with climbers.
8. Parents of teenagers **has / have** many worries and many joys.
9. **Does / Do** your brother and sister both have college degrees?
10. There **is / are** at least 28 moons around Jupiter.

QUIZ 8 Using Adjectives (Chart 6-8)

Directions: Underline the adjectives.

SITUATION: *Movie Night at Home*

Example: Let's stay home and watch a <u>new</u> movie.

1. We enjoy watching funny movies.
2. We have a big screen and good sound.
3. We can sit on our comfortable sofa.
4. I'll make some hot, buttery popcorn.
5. Some refreshing, cold drinks will taste great with the popcorn.
6. This movie is long. It will be late when it ends.

QUIZ 9 Using Adjectives (Chart 6-8)

Directions: Add the given adjectives to the sentences. Rewrite the complete sentence.

Example: old, busy   The man crossed the street.
   The old man crossed the busy street.

1. *small, new*   He used a laptop to check messages.
   _____

2. *clean, lower*   Nan put the clothes into the drawer.
   _____

3. *funny, friendly*   The story was about a monster.
   _____

4. *worried, angry*   The manager looked at the workers.
   _____

5. *happy, favorite*   The child played with her toy.
   _____

QUIZ 10 Using Nouns as Adjectives (Chart 6-9)

Directions: Correct the errors in adjectives.

Example: The vegetables at the ~~vegetables~~ vegetable market are always fresh.

1. Your flowers garden has many unusual flowers.

2. The mosquitos were really bad on our camping trip. I got a lot of mosquitos bites.

3. There is customers parking in front of the store. The customers are happy about that.

4. I see three spiders webs in the bathroom. I hate spiders!

5. All the computers printers in the library are new, but the computers are old.

6. Three people in our office are celebrating their birthdays tomorrow. There will be a lot of birthdays cake to eat.

7. Don't throw away the eggs cartons. We will put the hard-boiled eggs in them.

8. I love all the noodles in this soup. It's great noodles soup.

9. My doctor gave me an exercises plan. I have to do my exercises every day.

10. Collin lives in a two-bedrooms apartment. The bedrooms are quite large.

QUIZ 11 Personal Pronouns: Subjects and Objects (Chart 6-10)

Directions: Complete the sentences with the subject or object pronouns: **she, her, they, them, we, us,** or **it.** The first one is done for you.

SITUATION: *Anniversary Party*

Mr. and Mrs. Walters celebrated their fiftieth wedding anniversary. Their children had a party for _____them_____ . More than 100 people came to celebrate with
                                          1
_____ . _____ enjoyed sharing this special occasion with
         2   3
their friends.

SITUATION: *Twin Brothers*

My twin brother and I are very similar. _____ have the same friends and
                                                    4
enjoy the same activities. _____ wear the same style of clothes. Sometimes
                                    5
our friends and teachers mix _____ up, but my brother and I don't care.
                                        6
_____ have a good time joking with _____ .
         7   8

(continued on next page)

SITUATION: *Sara's Accident*

Sara got a new car last week. _____9_____ planned to drive to a friend's house, but on the way a tree branch fell on _____10_____. The roof of the car was damaged. Sara was sad but also very happy that the branch didn't hurt _____11_____.

QUIZ 12 Possessive Nouns   (Chart 6-11)

Directions: Use the correct possessive form of the given nouns to complete the sentences.

Example: earth   The ____*earth's*____ orbit around the sun takes 366 days in a leap year.

1. teachers   The _____ mailboxes are near their classrooms.
2. Carlos   Let's see if we can stay at _____ summer house during vacation.
3. wife   My _____ brother is doing research in the Sahara Desert.
4. children   The _____ names all start with "A"—Anna, Andy, and Alex.
5. hospitals   The _____ nurses don't want 12-hour workdays.
6. woman   It was hard to hear the _____ voice.
7. grandson   We will be attending our _____ graduation next week. We are so proud!
8. students   The teacher handed back the _____ papers. She was pleased with their work.
9. theater   The movie _____ sound system was terrible.
10. city   The _____ mayor promised to improve public transportation.

QUIZ 13 Who's vs. Whose (Chart 6-12)

A. Directions: Complete the questions with **Who's** or **Whose**.

SITUATION: *Questions About Your Family*

Example: ___Who's___ the oldest person in your family?

1. _____ your favorite relative?

2. _____ still living in your hometown?

3. _____ children are in college?

4. _____ not married yet?

5. _____ job is the most interesting?

B. Directions: Write complete questions using **Who's** or **Whose**. Use the same verb tense that is in the answer in parentheses.

SITUATION: *Our Neighbors*

Example: A: ___Who's the woman next door?___

B: Ms. Bell. (*Ms. Bell is the woman next door.*)

1. A: _____

   B: Beth. (*Beth is the new neighbor.*)

2. A: _____

   B: Mary's. (*Mary's dog always barks at the mailman.*)

3. A: _____

   B. The Johnsons'. (*The Johnsons' car is parked across the street.*)

4. A: _____

   B: Mr. Babcock. (*Mr. Babcock is washing windows.*)

5. A: _____

   B: Everyone. (*Everyone is coming to the barbecue.*)

QUIZ 14 Possessive Pronouns and Adjectives   (Chart 6-13)

Directions: Circle the correct words to complete the sentences.

Example: Austen forgot (his) / he's   cell phone at home.

1. Could I borrow a pen? **My / Mine** is out of ink.
2. Ingrid left **hers / her** laptop at school last night, and it was gone this morning.
3. A friend of **my / mine** brought flowers when I was in the hospital.
4. We don't agree. You have **your / yours** opinion, and we have **our / ours**.
5. The cat carried a mouse in **its / it's** mouth.
6. There are clouds in the sky, but **its / it's** not going to rain today.
7. Newlyweds Simone and Andre work at the same company in the same department. **Their / There / They're** desks are next to one another. **Their / There / They're** happy to be so close. Friends of **their / theirs** wonder about this, but Simone and Andre think **its / it's** good for **their / there / they're** relationship.
8. A: Look at this wonderful dessert! Kelly made it for **our / ours** party tonight.
   B: I know. **Its / It's** very rich. There are three kinds of chocolate in **it / its**.

QUIZ 15 Reflexive Pronouns   (Chart 6-14)

Directions: Complete the sentences with *myself, yourself, himself, herself, itself, ourselves, yourselves,* or *themselves*.

Example: You got 100 percent on the exam. You should be proud of _____*yourself*_____.

1. Luis works too much! He needs to take better care of _____.
2. You two have a good time at the party. Enjoy _____!
3. When Yoko was cutting vegetables, she accidentally cut _____.
4. My dream is to own my own company and work for _____.
5. When little Ella wakes up from her nap, she likes to play in her room and talk to _____.
6. Before we begin the meeting, let's all introduce _____.
7. Jin looked at _____ in the mirror and decided to comb his hair.
8. The cat is licking _____ clean.
9. The children played by _____ in the park while their mothers watched.
10. Tasha, you need to have more confidence. You need to believe in _____.

QUIZ 16 Another vs. The Other (Chart 6-15)

Directions: Circle the correct word(s) to complete the sentences.

Example: I have two brothers. One is named David and **another / (the other)** is named Michael.

1. I didn't understand that math problem. Could you show us **another / the other** one?

2. We tried two new dishes at the Chinese restaurant. I liked one a lot, but I didn't eat much of **another / the other**.

3. This is a dangerous intersection. Last night there was **another / the other** car accident here.

4. The boots I bought weren't very comfortable, so I took them back and got **another / the other** pair.

5. We can't decide what to name our baby. We especially like two names. One is Hannah, and **another / the other** is Rosemary. I like one name, and my husband likes **another / the other**.

6. Mario has experienced three earthquakes. One was in Turkey, **another / the other** was in Italy, and **another / the other** was in Japan.

7. We had a bad windstorm yesterday, and the weather forecast says there will be **another / the other** tomorrow.

8. This word has four definitions. I understand the first three definitions, but not **another / the other** one.

QUIZ 17 Other(s) vs. The Other(s) (Chart 6-16)

Directions: Complete the sentences with *other(s)* or *the other(s)*.

Example: Lena, Ann, and May are friends. Lena is in third grade, but ___the other___ girls are fourth graders.

1. I had five homework assignments for the weekend. I've finished two, but now I have to complete _____ before school tomorrow.

2. Juan speaks Portuguese and Spanish. Does he speak any _____ languages?

3. There is only one baked potato left. All of _____ were eaten at lunch.

4. Scientists have discovered a dinosaur bone in this area. They are looking for _____.

(continued on next page)

5. The morning flight to London is full, but _____ flights later in the day may have seats available.

6. Some animals have stripes, but _____ have spots.

7. There are four pears in the dish. Two of them aren't ripe yet, but _____ ones are good.

8. The Italian flag has three colors. One color is red, and _____ are green and white.

9. The teacher caught five students cheating. One student seemed embarrassed, but _____ didn't.

10. Plants need different environments to grow in. For example, some plants prefer shade and lots of water, while _____ grow best in full sun with little water.

QUIZ 18 Forms of *Other* (Charts 6-15 → 6-17)

Directions: Complete the sentences with *another*, *other*, *others*, *the other*, or *the others*.

Example: My favorite sport is tennis, but ___*another*___ sport I enjoy is badminton.

1. Let's order two pizzas. One will be vegetarian. What shall we have on _____ one?

2. Oh, no. Look! I found _____ gray hair on my head!

3. Of all the vegetables I planted this spring, only the peas did well. _____ died.

4. These jeans have a small hole in them. Do you have any _____ in this size?

5. The bus I was waiting for broke down, so I had to catch _____ one.

6. Some people need eight hours of sleep, but _____ need less.

7. I burned your hamburger. I'll make you _____ one.

8. The dog ran away with my only pair of dress shoes. I found one shoe in the garden and _____ under the stairs.

9. You have one idea about how to solve the problem, and I have _____. Since we can't agree, let's keep looking for _____ ideas.

QUIZ 19 Chapter Review

Directions: Correct the errors.

Example: My son lost two ~~tooths~~ *teeth* last year.

1. There are 30 day in the month of April.

2. Our apartments manager is out of town this week.

3. Who's glasses are on your desk?

4. She was born in September 8, 2003.

5. Mr. Lee company recycles old computers.

6. The cars in the city makes the streets very noisy.

7. Many people are waiting in the rain to buy tickets for the concert. Everyone seem patient, but cold.

8. The childrens swimming pool in the city is open to all children.

9. I broke my right hand, so I need to write with the another one.

10. The dancers practiced in the studio all morning their dance steps.

CHAPTER 6 – TEST 1

Part A *Directions:* Circle the correct completions.

1. I can't change my **phone / phones / phone's** password. The **instructions / instruction's / instructions'** aren't clear.

2. I have two children. My **daughter / daughter's / daughters'** **name / names / name's** are Emma and Ellen.

3. **Today / Today's / Todays** newspaper has several **article / articles / articles'** about the growth of our city.

4. **Beth / Beth's / Beths'** computer is similar to mine. Both of our **computer / computers / computer's** are new and fast.

5. People at the lecture thought the **speaker / speaker's / speakers** **idea / ideas / ideas'** were interesting.

Part B *Directions:* Circle the correct completions.

1. A: Which notebook is **your / yours**?

   B: **My / Mine** is the one with the star on the cover.

2. Toshi is proud of **his / him** daughters. They did well in school, and now each one has **she / her** own company and works for **herself / herselves**.

3. A: Are these sunglasses **your / yours**?

   B: No, they're **hers / she's**.

4. The Bakers invited Graciela and **I / me** to a piano concert in the city. **Their / There / They're** going to pick us up in **their / there / they're** car and drive us **their / there / they're**. **It's / Its** going to be fun.

5. The city bus was so full that it couldn't pick up my brother and **I / me**. **We / Us** had to wait an hour for the next bus.

Part C *Directions:* Complete the sentences with *another, other, others, the other,* or *the others.*

1. Alex found several colorful shells on the beach. He brought one home, but left _____ on the sand.

2. After the Blakes got a puppy, they decided they needed _____ one so it would have a playmate.

3. We're sorry we can't attend your party. We have _____ plans.

4. Some people prefer traveling by train, but _____ prefer driving.

5. Three of my friends have unusual pets. One has an iguana. _____ has a python snake. _____ one has a baby alligator.

Part D *Directions:* Circle the correct sentence in each pair.

1. a. Tony was born in April 2, 1998.
   b. Tony was born on April 2, 1998.

2. a. I like to take in the morning in the park long walks.
   b. I like to take long walks in the park in the morning.

3. a. Whose phone is ringing?
   b. Who's phone is ringing?

4. a. Every students in the class are working hard and making progress.
   b. Every student in the class is working hard and making progress.

5. a. Several language schools offer university preparation.
   b. Several languages schools offer university's preparation.

CHAPTER 6 – TEST 2

Part A *Directions:* Circle the correct completions.

1. My **friend / friend's / friends'** dog had three **puppy / puppies / puppy's** last night. They are so cute!

2. Professor **Brown / Browns / Brown's** math **class / classes / classes'** are very difficult.

3. My sister just had twin **boy / boys / boys'**. Her **babys / babies / babies'** names are Tyler and Spencer.

4. **Student's / Students' / Students** cannot register for classes at this time. The **university / university's / universities** main computer is having technical **problem / problem's / problems**.

Part B *Directions:* Circle the correct completions.

1. Did you know that the Smiths are cousins of **me / mine**?

2. **Your / Yours** hand is bleeding. How did you cut **you / yourself**?

3. The baby bird is chirping in **its / it's** nest, waiting for **its / it's** mother. **Its / It's** hungry.

4. My husband works at home. **His / Him** office is in the basement. **He / His** built it **himself / itself**. Having a quiet work space makes **he / him** more productive now. **His / Him** manager is happy with this, too.

5. Luis and **I / me** are vacationing in Mexico next month. We are staying at **his / he's** parents' house in Mazatlán.

6. A: Whose seats are these?

   B: Those people over **their / there / they're**. **Their / There / They're** waiting in line for popcorn.

Part C *Directions:* Complete the sentences with *another, other, others, the other,* or *the others.*

1. My school offers instruction in several different sports. Golf is one, and swimming is _____.

2. So far, your suggestions have been very helpful. What _____ ideas do you have?

3. If Matt gets _____ failing grade, he'll have to repeat the class.

4. My friends have given me three surprise birthday parties in the past. I was surprised by the first, but not by _____.

5. Some people like to get up early, and _____ like to sleep until noon.

6. The window washers only need to wash the windows on the front of the building. _____ windows are clean.

Part D *Directions:* Circle the correct sentence in each pair.

1. a. Nancy flower's garden has many rose in it.
   b. Nancy's flower garden has many roses in it.

2. a. We rented for a month a cabin in the mountains.
   b. We rented a cabin in the mountains for one month.

3. a. There are several cars in our driveway. Who do they belong to?
   b. There are several cars in our driveway. Who's are they?

4. a. My sister's husband is a really funny guy. Him always tells jokes.
   b. My sister's husband is a really funny guy. He always tells jokes.

5. a. The apples in the box were rotten, so we didn't eat them.
   b. The apple in the box were rotten, so we didn't eat it.

CHAPTER 7 Modal Auxiliaries, the Imperative, Making Suggestions, Stating Preferences

QUIZ 1 The Form of Modal Auxiliaries (Chart 7-1)

Directions: Circle the correct words. Pay attention to modal forms.

1. Stan   **ables to ride / is able to ride**   his bike ten miles to work every day.
2. The delivery driver   **couldn't find / couldn't finding**   our house.
3. John isn't here yet. He   **may be / maybe**   late because of traffic.
4. Shelley   **cans help / can help**   us if she has time.
5. You   **may borrow / may borrows**   my car if you are back by 6:00 p.m.
6. Would you   **to come / come**   to my party on the weekend?
7. You   **have to try / has to try**   this dessert. It's delicious.
8. Mr. Daly   **will not buy / won't buying**   a new car next year.
9. We   **aren't able to eat / can't to eat**   at that restaurant. It's too busy.
10. Do we   **have leave / have to leave**   right now?

QUIZ 2 Expressing Ability (Chart 7-2)

A. *Can* and *Can't*

Directions: Complete the sentences with **can** or **can't**.

Example: Bees __can__ see many colors, but they __can't__ see the color red.

1. Newborn babies _____ walk, but newborn horses _____.
2. Dogs _____ talk to people, but they _____ bark.
3. Young children _____ ride in cars, but they _____ drive them.
4. I _____ check email on my smartwatch, but I _____ take pictures with it.
5. Spiders _____ make webs, but they _____ make honey.

(continued on next page)

B. Could and Be Able To

Directions: Complete the sentences with the correct form of *could, couldn't, be able to,* or *not be able to.* More than one answer may be correct.

Example: Three years ago, Jean-Pierre __couldn't / wasn't able to__ understand spoken English, but now he can.

1. When I was a teenager, I _____ eat a whole pizza by myself, but I can't now.

2. Jacob _____ speak Spanish, but his mother could speak it fluently.

3. My dad called me online. I _____ hear him fine, but I _____ see him because my webcam didn't work.

4. When astronaut Scott Kelly lived on the International Space Station, he _____ see the earth and the moon through the window.

QUIZ 3 Possibility vs. Permission (Chart 7-3)

Directions: Decide if the modal verbs in *italics* express possibility or permission. The first one is done for you.

	permission	possibility
1. We *may have* hot chocolate later.		✓
2. You *may leave* now.		
3. I *might need* some help.		
4. You *can go* home early.		
5. I *may need* extra time.		
6. You *might not get* a seat in the theater.		
7. You *may not sit* there.		
8. You *can't eat* that candy.		
9. It *may be* sunny tomorrow.		
10. We *can begin* work tomorrow.		
11. The bus *may be* late.		

QUIZ 4 Expressing Possibility (Chart 7-3)

Directions: Rewrite the sentences with the words in parentheses.

Example: It might rain tomorrow. (*maybe*)

    Maybe it will rain tomorrow.

1. Maybe we'll go away this weekend. (*might*)

2. It might snow tomorrow. (*maybe*)

3. Our basketball team may win the championship. (*maybe*)

4. Joan may be in the hospital. (*might*)

5. David might take the driving test tomorrow. (*may*)

6. Maybe Sara will go shopping with us this afternoon. (*might*)

7. James may be late for the meeting. (*maybe*)

8. Maybe my keys are in my backpack. (*may*)

9. We might watch a movie tonight. (*may*)

10. The car might be ready later this afternoon. (*maybe*)

QUIZ 5 Meanings of Could (Charts 7-2 and 7-4)

Directions: Decide if *could* expresses past, present, or future. Then decide if it expresses ability or possibility. The first one is done for you.

	past	present	future	ability	possibility
1. Just think! You *could* graduate in one year!			✓		✓
2. I *could* touch my toes when I was younger.					
3. Traffic is a little heavy. There *could* be an accident.					
4. Mr. Chu *could* speak fluent English a few years ago, but he's forgotten some now.					
5. You *could* help me with the dishes after dinner.					
6. There *could* be some good sales at the mall today.					
7. I *could* be at work until late tonight. I have to finish a project.					
8. Katherine *could* write her name when she was three years old.					
9. We *could* go on a picnic this afternoon if the weather is nice.					
10. Where *could* my keys be? They're not in my purse or my pocket.					
11. Claire *could* be a professional actress someday. She's very good!					

QUIZ 6 Ability, Possibility, and Permission (Charts 7-2 → 7-4)

Directions: Circle the word or phrase that has the same meaning as the words in *italics*.

Example: **It's possible** Rick is at the grocery store, or he can / (could) be studying at the library.

1. *It's possible* we'll have a storm tomorrow. In fact, it **maybe / might** snow.

2. Josh is a really good cook. He *is able to* cook Italian food and **can / may** prepare a few Indian dishes, too.

3. A: I need to go to the bookstore before it closes, but I'll never get there in time!

    B: You *have* my *permission* to use the car. You **could / can** drive there. It's no problem!

4. Martin looks pale. *It's possible* he's sick. He **may / can** have a fever.

5. Kathy *doesn't have permission* to stay out late. She **could not / may not** be out past 10 P.M.

6. Mr. White **may / could** jump very high when he was young. He *was able to* leap over a fence easily.

7. Ben *isn't able to* find his wallet, so he **maybe not / can't** pay for his dinner.

8. A: Do you *have* your parents' *permission* to go to the movies with us?

    B: Yes, they said I **can / maybe** go with you.

9. The test was easy. *It's possible* I got a high grade. **Could / Maybe** I got 100 percent!

10. It *was not possible* for me meet you yesterday. I was sick and **couldn't / can't** go out.

QUIZ 7 Polite Requests (Charts 7-5 and 7-6)

Directions: Check (✓) all the correct modals for each polite request.

Example: A: Do you understand how this computer program works?

B: Sort of, but not really. ___ you explain it to me one more time, please?

___ May   ✓ Could   ✓ Can   ✓ Would   ✓ Will

1. A: Mr. Andrews, I have a headache. ___ I leave early, please?
   B: Of course, Ms. Ochoa. I hope you feel better tomorrow.

   ___ May   ___ Could   ___ Can   ___ Would   ___ Will

2. A: Mom, I finished my homework. ___ you check it?
   B: Sure. I'd be happy to.

   ___ May   ___ Could   ___ Can   ___ Would   ___ Will

3. A: The copy machine is out of paper. ___ you please refill it?
   B: Yes. I'll take care of it in a few minutes.

   ___ May   ___ Could   ___ Can   ___ Would   ___ Will

4. A: Hi, Mr. Martin. It looks like you have your hands full. ___ I help you carry your groceries?
   B: That would be great, Jim. Thanks a lot!

   ___ May   ___ Could   ___ Can   ___ Would   ___ Will

5. A: This rice is delicious. ___ I have another bowl?
   B: Of course! I'm so happy you like it.

   ___ May   ___ Could   ___ Can   ___ Would   ___ Will

6. A: Excuse me. I'm looking for a coffee shop. ___ you tell me if there's one nearby?
   B: Sure. There's one just around this corner.

   ___ May   ___ Could   ___ Can   ___ Would   ___ Will

7. A: Good morning. ___ I help you find something in the store?
   B: No thank you. I'm just looking around.

   ___ May   ___ Could   ___ Can   ___ Would   ___ Will

8. A: Mikael, ___ you hand me the phone, please?
   B: Sure, Dad.

   ___ may   ___ could   ___ can   ___ would   ___ will

9. A: Excuse me. ___ I sit here?
   B: I'm sorry, but this seat is already taken. My friend is coming right back.

   ___ May   ___ Could   ___ Can   ___ Would   ___ Will

10. A: ___ you please tell me what happened?
    B: No. I'm too embarrassed to talk about it.

    ___ May   ___ Could   ___ Can   ___ Would   ___ Will

QUIZ 8 Should and Ought To (Chart 7-7)

A. Directions: Rewrite the sentences using **should** or **shouldn't** and a verb.

SITUATION: *Traveling to a Different Country*

Example: It's a good idea to try learn about the culture.

Travelers ___should try___ to learn about the culture.

1. It's a good idea to be careful with your passport and other travel documents.

   You _____ careful with your passport and other travel documents.

2. It's a good idea to learn to say "please" and "thank you" in the local language.

   People _____ to say "please" and "thank you" in the local language.

3. It's not a good idea to expect everything to be the same as at home.

   Travelers _____ everything to be the same as at home.

4. It's a bad idea to be impatient if someone doesn't understand you.

   You _____ impatient if someone doesn't understand you.

5. It's a good idea to try food that is different from food in your country.

   You _____ food that is different from food in your country.

B. Directions: Read the problem. Give advice using **should** or **ought to**. More than one answer is possible.

Example: A: I'm sleepy.

   B: ___You should / ought to go to bed early tonight.___

1. A: My computer screen keeps crashing.

   B: _____

2. A: My foot hurts. It's painful when I walk.

   B: _____

3. A: I'm so hungry. I didn't have time for breakfast.

   B: _____

4. A: Henry is so depressed. He failed his chemistry test.

   B: _____

5. A: My coffee is cold.

   B: _____

QUIZ 9 *Should / Ought To vs. Had Better* (Charts 7-7 and 7-8)

A. Directions: Complete the sentences using *should/ought to* or *had better*. Use the negative form where appropriate.

Examples: The boys ___had better___ not play soccer in the house. They might break something.

You ___should___ try on this hat. It's cute.

1. You _____ leave early for the airport. That way you won't have to rush.

2. You _____ lose your passport. You won't be able to travel.

3. We _____ put away the game. It's almost time for bed.

4. A: My horse's leg is bleeding.

   B: Really? You _____ call the vet. It could be something serious.

5. Children _____ spend too much time online.

B. Directions: Make suggestions with *should, ought to,* or *had better.* More than one answer may be correct.

Examples: Billy is walking on the carpet with wet shoes. His mother is upset.

Billy ___should take off his shoes___. OR

Billy ___ought to be more careful___. OR

Billy ___had better apologize to his mother___.

1. You just touched a hot pan and burned your finger.

   I _____.

2. There is a snake in front of your door, and you need to go out.

   I _____.

3. It's raining, and there's a leak in Suzette's roof. Water is coming into her apartment.

   She _____.

4. Mr. Sato is shopping at the market and discovers that he has no money with him.

   He _____.

5. I just tripped on your stairs and almost fell! One of the steps is broken.

   You _____.

QUIZ 10 Necessity: *Have To, Have Got To,* and *Must* (Chart 7-9)

Directions: Circle the correct modals.

Example: Tomorrow I **(have to)/ had to** go to the dentist.

SITUATION: *In Class*

1. Yesterday, our teacher **must / had to** leave school early.
2. We **had got to / had to** work quietly by ourselves until the end of class.
3. Why did she **have to / had to** leave early?

SITUATION: *At the Airport*

4. Everyone **must / have to** have a ticket to go to the gate.
5. Sometimes we **has to / have got to** wait in a long line.
6. Did you **have to / had to** check your bags?

SITUATION: *Going to a Party*

7. I **have got to / had to** get ready for the party. It starts in an hour.
8. We **must / had to** be on time. I hate to be late!
9. Last weekend, I **have to / had to** buy a new dress to wear.
10. My roommate **has got to / have got to** lock the door when we leave.

QUIZ 11 *Do Not Have To* and *Must Not* (Chart 7-10)

Directions: Complete the sentences with **don't have to, doesn't have to,** or **must not**.

SITUATION: *Using the Internet*

Example: Students ____must not____ surf the internet when they are in class.

1. You _____ be an expert to use the internet effectively.
2. People _____ download music or movies illegally.
3. Sometimes you can get music or movies for free. You _____ pay for them.
4. Be careful! You _____ give out too much personal information.
5. The Wi-Fi at the library is so fast! I _____ wait long for pages to load.
6. On some blogs, you _____ be a member. Anyone can write on them.

(continued on next page)

7. My sister _____ use a computer to check her email; she uses her cell phone.

8. When people post messages online, they _____ be so careful with grammar.

9. You _____ believe everything you read online. Some of it isn't true!

10. Parents _____ allow their children to spend too much time online.

QUIZ 12 Review of *Must, Have To,* and *Had To* (Charts 7-9 and 7-10)

Directions: Complete the sentences with a modal and the given verb. Use ***must, have to, had to, must not, don't have to,*** or ***didn't have to.*** More than one answer may be correct.

SITUATION: *Vacation Travel*

Example: plan   I want to travel in Europe. I ___*have to plan*___ everything carefully.

1. ask   Before Sam goes on vacation, he _____ his boss for time off from work.

2. have   If you travel to a different country, you _____ a passport.

3. forget   You _____ your passport! You can't travel to a different country without it.

4. worry   If you stay in your own country, you _____ about your passport that much.

5. go   Two years ago someone stole my wallet. I _____ to the police for help.

6. take   Last December, Maggie flew to Hawaii. She _____ any warm clothes with her.

7. buy   People _____ plane tickets several months before their trip to get the best price.

8. pay   My sister works for an airline. She _____ full price for flight tickets.

9. be   When Erik travelled to Japan, he _____ change his dollars to yen.

10. travel   When I was a kid, I _____ with my parents, but now I can travel alone.

QUIZ 13 Logical Conclusion vs. Necessity (Charts 7-9 → 7-11)

Directions: Decide if *must* expresses logical conclusion or necessity. The first one is done for you.

	logical conclusion	necessity
1. You *must* be Faisal's son. You look just like your father.	✓	
2. You *must* wash your hands with warm water and soap.		
3. Ellen *must* know about the accident. She's a police officer.		
4. You look really tired. You *must* take a break.		
5. We *must* feed our pets before we leave for work.		
6. Julie *must* like school. She's ready an hour early every morning.		
7. I see lightning in the distance. There *must* be a storm coming.		
8. Children, you *must* stay in your seats until the bell rings.		
9. You got a high score on the exam! You *must* feel wonderful.		
10. Your grades are low. You *must* study more.		
11. I'm sorry that your mom is so sick. You *must* be worried.		

QUIZ 14 Making Logical Conclusions: Must (Chart 7-11)

Directions: Complete the conversations with *must* or *must not*.

Example: A: Jose isn't here yet.

B: The traffic _____must_____ be really bad today.

1. A: Wow! You're shaking! You _____ be cold.

   B: I am. I hope I'm not getting sick.

2. A: Sue has tried on three dresses already, but she isn't going to buy any of them.

   B: She _____ like them.

3. A: A car just parked in front of our house.

   B: That _____ be Rudy. He's picking me up.

4. A: Bill and Alicia's son is getting married next month.

   B: They _____ be very excited about the wedding!

5. A: Lucy misses class at least twice a week.

   B: She _____ care about getting a good grade.

(continued on next page)

6. A: I am trying to reach Josh on his cell phone, but he won't answer.

   B: He usually answers his phone. He _____ have it with him today.

7. A: There are a lot of bees flying around outside our bedroom window.

   B: Uh-oh! There _____ be a beehive nearby.

8. A: Daniel and Yuko are leaving for Tokyo on Saturday.

   B: They _____ have a lot of packing to do.

9. A: The cat won't eat the fish I gave her.

   B: It _____ like the fish.

10. A: Nancy practices piano for two hours every day.

    B: She _____ be serious about becoming a professional musician.

QUIZ 15 Tag Questions with Modals   (Chart 7-12)

Directions: Complete the tag questions.

Example: Derek can speak Turkish, _____*can't*_____ he?

1. You have to study for semester exams, _____ you?
2. Children should get lots of fresh air and exercise, _____ they?
3. Liz can't cook Vietnamese food, _____ she?
4. He doesn't have to work on the weekend, _____ he?
5. You will help me with my project, _____ you?
6. I shouldn't worry so much, _____ I?
7. Cathy and Bev would rather stay home, _____ they?
8. We'll be late for school today, _____ we?
9. You couldn't find the website, _____ you?
10. Patrick wouldn't like this movie, _____ he?

QUIZ 16 Imperative Sentences: Giving Instructions (Chart 7-13)

Directions: Complete the sentences with the imperative.

SITUATION: *Baking a Cake*

Example: You have to buy flour, sugar, butter, and eggs.

_____Buy_____ flour, sugar, butter, and eggs.

1. You need to prepare a cake pan.

   _____ a cake pan.

2. It's a good idea to measure flour, sugar, and butter carefully.

   _____ flour, sugar, and butter carefully.

3. You have to mix the butter, sugar, eggs, and milk.

   _____ the butter, sugar, eggs, and milk.

4. After that, you should carefully add the flour, baking powder, and salt.

   _____ the flour, baking powder, and salt.

5. Then you need to pour the cake batter into the pan.

   _____ the cake batter into the pan.

6. You have to bake the cake for 25 to 30 minutes.

   _____ it for 25 to 30 minutes.

7. After 25 minutes, you should test the cake to see if it's done.

   After 25 minutes, _____ the cake to see if it's done.

8. When the cake is done, you have to let it cool for ten minutes.

   When the cake is done, _____ it cool for ten minutes.

9. After that, you can take it out of the pan.

   After that, _____ it out of the pan.

10. When the cake is completely cool, you can decorate it.

    When the cake is completely cool, _____ it.

QUIZ 17 Making Suggestions: *Let's* and *Why Don't* (Chart 7-14)

Directions: Check (✓) the parts in the conversation that include suggestions. The first one is done for you.

SITUATION: *Plans After Work*

1. ✓ A: Why don't we meet at Cory's Café after work?
2. ___ B: That sounds nice. Let me finish this report first.
3. ___ A: Of course! Let's meet at about 6:00.
4. ___ B: I can't leave until about 6:15. Why don't you go early and get a table?
5. ___ A: OK. That sounds good.
6. ___ B: Let's have dessert there. I love their sweets.
7. ___ A: Yeah! Why don't we have some coffee and chocolate cake?
8. ___ B: Why do you want coffee? Isn't it a little late for coffee?
9. ___ A: It won't bother me. Why don't you have herbal tea instead?
10. ___ B: Good idea. They have many types to choose from.
11. ___ A: Great. Let's finish our work so we can go!

QUIZ 18 Prefer, Like ... Better, and Would Rather (Chart 7-15)

Directions: Complete the sentences with the correct form of *prefer*, *like*, or *would rather*.

Example: Kelly _____prefers_____ a laptop to a desktop computer.

1. Dr. Quinn _____ see her patients in the morning than in the evening.
2. Joanne _____ smaller cars better than bigger ones.
3. The Petersons _____ warmer climates to cooler ones.
4. I _____ watch a movie at home than go out to a theater.
5. Monica _____ fish to red meat.
6. Harry _____ read the news online than in the newspaper.
7. My father _____ books to movies.
8. Scott and Jeff _____ to swim in the sea better than in the lake.
9. I _____ pens to pencils.
10. We _____ spend our vacations with our grandchildren than travel.

QUIZ 19 Chapter Review

Directions: Circle the correct completions. The first one is done for you.

1. I'm not feeling well. ___ I lie down on your sofa?

    a. Could b. Would c. Must

2. You ___ drive more slowly or you're going to get a speeding ticket.

    a. could b. ought to c. had to

3. ___ I borrow your pencil? Mine just broke.

    a. Would b. Ought c. May

4. You ___ write to me when you get to Paris. Just text me a photo.

    a. must not b. don't have to c. couldn't

5. Jessica doesn't look healthy. She ___ see a doctor.

    a. should b. can c. would

6. ___ you help me lift this box? It's too heavy.

    a. Will b. May c. Should

7. I just heard a car door. Dad ___ be home.

    a. would b. can c. must

8. Our dog ___ count! Listen to him bark as I say the numbers.

    a. should b. can c. must

9. You ___ be Rita. Mary told me you would be here today.

    a. can b. will c. must

10. ___ you still be here when I get back?

    a. Will b. Would c. May

11. What can we do about Grandma? She ___ drive anymore. She's dangerous.

    a. doesn't have to b. might not c. shouldn't

12. You ___ pay the electric bill. I already paid it online.

    a. don't have to b. shouldn't c. can't

13. ___ you turn down the heat, please?

    a. Could b. Should c. May

(continued on next page)

14. When Tara was two years old, she ____ recognize letters and numbers.

    a. could   b. should   c. might

15. We ____ use our credit card at the store because their computers weren't working.

    a. shouldn't   b. wouldn't   c. couldn't

16. ____ you please repeat that?

    a. Could   b. May   c. Should

17. You have worked late every night this week! You ____ be really tired.

    a. ought to   b. must   c. will

18. You ____ stop playing that game right now. You have to study, or you won't be ready for your exam tomorrow.

    a. had better   b. don't need to   c. would

19. I'm bored. ____ to the mall?

    a. Let's go   b. Why don't we go c. Go

20. ____ we will come home from our vacation early. We haven't decided yet.

    a. May be   b. Might   c. Maybe

21. Jim never drinks coffee. He ____ like it.

    a. should not   b. could not   c. must not

CHAPTER 7 – TEST 1

Part A *Directions:* Circle the correct completions.

1. A: Ouch! I cut my hand!

    B: You ____ clean it well and put a bandage on it.

    a. will　　　　b. had better　　　c. may

2. A: Is this your eraser?

    B: No. It ____ be Diana's. She was sitting at that desk.

    a. had better　　　b. will　　　c. must

3. A: Let's go to a movie this evening.

    B: That sounds like fun, but I can't. I ____ finish this essay before I go to bed tonight.

    a. have got to　　　b. would rather　　　c. ought to

4. A: We ____ hurry. The store closes in 30 minutes.

    B: I'll finish my shopping quickly.

    a. should　　　b. will　　　c. may

5. A: I did it! I did it! I got my driver's license!

    B: Congratulations, Michelle. I'm really proud of you.

    A: Thanks, Dad. Now ____ I have the car tonight? Please?

    B: No. You're not ready for that quite yet.

    a. will　　　b. should　　　c. may

6. A: What do you want to do on our vacation this summer?

    B: I ____ go camping than spend time in a big city.

    a. could　　　b. would rather　　　c. prefer

7. A: Are you going to the conference in Atlanta next month?

    B: I ____. I'm not sure yet.

    a. will　　　b. have to　　　c. might

8. A: What shall we do after the meeting this evening?

    B: ____ pick Jan up and go out to dinner together.

    a. Why don't　　　b. Let's　　　c. Should

(continued on next page)

9. A: Have you seen my denim jacket? I ____ find it.

   B: Look in the hall closet.

   a. may not   b. won't   c. can't

10. A: Bye, Mom! I'm going out to play soccer with my friends.

    B: Wait a minute, young man! You ____ clean your room first.

    a. had better not   b. have to   c. would rather

Part B *Directions:* Complete the sentences with **can, could,** or **might.** More than one answer may be correct.

1. Look at little Ben! He _____ walk!

2. I don't know where Irina is. She _____ be shopping.

3. When I was younger, I was in great shape. I _____ run for an hour.

4. Turn on the TV. There _____ be news about the accident.

5. Hannah _____ play the piano when she was twelve, but now she can't.

Part C *Directions:* Answer the questions. Write complete sentences. Use the modal in *italics* in your answer.

1. What is something you **had to** do yesterday?

   _____

2. What is something you **must** do before you can drive a car?

   _____

3. What is something that children **shouldn't** do?

   _____

4. What is something you **should** do when you don't feel well?

   _____

5. What is something you **have to** do every morning?

   _____

Part D *Directions:* Correct the verb errors.

1. I'm feeling hot. I ought take my temperature.

2. Would I borrow your pen? Mine isn't working.

3. I don't feel like cooking. Let's we order a pizza.

4. We'll be free on Saturday. We could to meet then.

5. Look at the sky. It could snow tomorrow, could it?

6. Thomas is late. He can have car trouble again.

7. Children don't have to play with matches. They can start fires.

8. Why we don't go for a walk after dinner? It's such a nice evening.

9. I don't want to stay home this weekend. I rather go hiking.

10. Jenny have to be more careful with her glasses. She has broken them twice.

CHAPTER 7 – TEST 2

Part A *Directions:* Circle the correct completions.

1. A: Do you think Majid will quit his job?

    B: I don't know. He ____. He'll wait a few weeks to decide.

    a. must   b. may   c. will

2. A: The swimming pool has towels. You ____ bring one.

    B: Okay. I'll just bring my swimsuit.

    a. don't have to   b. must not   c. couldn't

3. A: I heard that Bill was seriously ill.

    B: Really? Well, he ____ be sick now. I just saw him riding his bike to work.

    a. doesn't have to   b. won't   c. must not

4. A: Henry should be here soon, ____ he?

    B: Yes. His plane arrives in ten minutes.

    a. should   b. shouldn't   c. doesn't

5. A: Did you climb to the top of the Statue of Liberty when you were in New York?

    B: No, I didn't. My knee was very sore, and I ____ climb all those stairs.

    a. might not   b. couldn't   c. must not

6. A: Rick, ____ work for me this evening? I'll take your shift tomorrow.

    B: No problem. I can work tonight instead of tomorrow.

    a. should you   b. would you   c. do you have to

7. A: What do you want to do with the dog when we are out of town?

    B: ____ we ask Alex to take care of him?

    a. Why don't   b. Let's   c. Will

8. A: You ____ attend the meeting tomorrow morning. It's important.

    B: OK. I'll be there.

    a. could   b. maybe   c. must

9. A: I have to pay bills today.

    B: That's good. You ____ forget to pay your credit card bill, or you will have a late fee.

    a. mustn't   b. couldn't   c. don't have to

10. A: Don't wait for me. I ____ late.

    B: OK.

    a. maybe   b. may to be   c. may be

Modal Auxiliaries, the Imperative, Making Suggestions, Stating Preferences

Part B *Directions:* Complete the sentences with *can*, *could*, or *might*. More than one answer may be correct.

1. These new glasses are much better. I _____ see much more clearly.
2. When I lived in Paris, I _____ speak French really well.
3. I don't know why the Martins aren't here. They _____ have another party to go to.
4. I don't know if we have enough help. We _____ need to call more people.
5. When Dr. Kim was in medical school, he _____ work for days with very little sleep.

Part C *Directions:* Answer the questions. Write complete sentences. Use the modal in *italics* in your answer.

1. What is something you **must** do today?
   _____

2. What is something you **should** do when you are at school?
   _____

3. What is something that people **shouldn't** do?
   _____

4. What is something you **have to** do before you go to bed tonight?
   _____

5. What is something you **had to** do last week?
   _____

Part D *Directions:* Correct the verb errors.

1. My grades are low. I had to better study more.
2. May you please open the window? It's hot in here.
3. I want to stay home tonight. Let invite some friends over.
4. We can't to come to your party. We will be out of town.
5. Susan maybe has a solution to the problem.
6. Jackie isn't here. She can be at home in bed.
7. You mustn't to walk in mud puddles.
8. Why we don't go out for dinner tonight?
9. You have to study tonight, haven't you?
10. I need to make a call. Would I borrow your phone for a minute?

CHAPTER 8 Connecting Ideas: Punctuation and Meaning

QUIZ 1 Punctuating with Commas and Periods (Chart 8-1)

Directions: Add commas and periods where appropriate. Capitalize as necessary.

SITUATION: *Beth's Party*

Example: Beth had a party with her friends**,** neighbors**,** and roommates**.**

1. Beth planned to serve pizza green salad and ice cream at the party
2. Sam Jeff and Ellen helped with the decorations Bob picked up the pizza and drinks
3. The party started at 7:00 P.M. several guests were late
4. A few people talked others played games and several people danced
5. Everyone had a wonderful time no one wanted to go home
6. Beth thanked everyone for coming and promised to have another party then she told everyone good night

QUIZ 2 Connecting Ideas: *And, But, and Or* (Charts 8-1 and 8-2)

Directions: Complete the conversations with ***and, but,*** or ***or.*** Add commas as necessary. The first one is done for you.

1. A: Do you want to watch a comedy __*or*__ an action movie?

   B: Let's watch a comedy. Jan, Cathy, _____ I saw an action movie last weekend.

2. A: I emailed Pierre two days ago _____ he hasn't answered.

   B: Did you write to his work email _____ his home email?

   A: His work email.

3. A: What do you like to do for fun?

   B: My favorite hobbies are gardening, bicycling _____ bird watching. How about you?

   A: I enjoy playing golf _____ it is an expensive sport. I don't play often.

   B: My favorite sports are soccer _____ swimming. They don't cost much.

(continued on next page)

4. A: Nicky, you can have spaghetti _____ a grilled cheese sandwich. Which do you want?

   B: I'd like a grilled cheese sandwich _____ a glass of milk, please.

5. A: How was your camping trip?

   B: Too cold! We slept in a tent _____ we got cold in the middle of the night. Then I tried to start a campfire _____ the wood was too wet. It didn't burn.

QUIZ 3 So vs. But   (Charts 8-2 and 8-3)

Directions: Complete the sentences with **so** or **but**.

Example: Norma likes soccer, __but__ I don't.

1. a. The soccer game was on TV, _____ I didn't watch it.

   b. The basketball game wasn't on TV, _____ I was able to watch it online.

   c. My favorite team won the game, _____ I was very happy.

2. a. My father is sick, _____ I will stay with him for a few days.

   b. My father is sick, _____ he doesn't want me to stay with him.

   c. The doctor gave my father some medicine, _____ he may feel better soon.

3. a. I made some chicken soup last night, _____ it wasn't very good.

   b. I forgot to put garlic and salt in the soup, _____ it didn't have much flavor.

4. a. Renee wants to go shopping, _____ she doesn't have any money.

   b. I usually use my credit card when I shop, _____ Renee prefers to pay cash.

QUIZ 4 Using Auxiliary Verbs After But   (Chart 8-4)

Directions: Complete the sentences with the correct auxiliary verbs.

Example: The peas in my garden are ripe, but the corn __isn't__ .

1. Sara wants to buy a new car, but her husband _____ .

2. Fred isn't ready, but I _____ .

3. Khaled has finished his test, but the other students _____ .

4. The children don't want to go to bed, but I _____ .

5. These lamps have energy-saving bulbs, but those lamps _____ .

6. Julie hasn't read that book, but most of her friends _____ .

(continued on next page)

7. I am not going to go out for lunch, but a few coworkers _____.

8. You won't have to wait for the next flight, but other people _____.

9. I got a text message from my dad, but my brother and sister _____.

10. Most of our class will graduate this year, but a few students _____.

QUIZ 5 Auxiliary Verbs After *And* and *But* (Charts 8-4 and 8-5)

Directions: Read the paragraph about the Sweet Family. Then complete the sentences with the correct auxiliary verbs. The first one is done for you.

The Sweet family has five people: the parents, two girls, and one boy. Natalie, the mom, owns a coffee shop, and the dad, Will, is a carpenter. They work while their kids are at school. The girls are Laura and Janey. Their brother's name is Daniel. They are students at Ravenswood High School. They love music and sing in the choir. Laura and Daniel are also in the Drama Club. Janey isn't in the Drama Club. She likes painting and drawing, so she is in the Art Club. Natalie and Will are proud of their children.

1. Natalie works, and so ____*does*____ Will.

2. Natalie works at her coffee shop, but Will _____.

3. Natalie has three kids, and Will _____ too.

4. Laura is a girl, and so _____ Janey.

5. Janey is in high school, and Laura and Daniel _____ too.

6. Laura and Janey sing in the choir, and Daniel _____ too.

7. Daniel likes music, and so _____ Laura and Janey.

8. Laura and Daniel are in the Drama Club, but Janey _____.

9. Janey likes drawing and painting, but Daniel and Laura _____.

10. Janey is in the Art Club, but Laura and Daniel _____.

11. Natalie is proud of the kids, and so _____ Will.

QUIZ 6 Using *And* + *Too, So, Either,* and *Neither* (Chart 8-5)

Directions: Read the information about two brothers, Joe and Sam. Complete the sentences using *too, so, either,* or *neither* and the correct auxiliary verb. Make true statements. The first one is done for you.

Joe ...
is 35 years old
is a math teacher
is married
has two daughters
lives in Los Angeles
is a vegetarian
has been to London
is learning French

Sam ...
is 33 years old
is a health club manager
is married
has two daughters
lives in San Francisco
is a vegetarian
has been to London
is learning Chinese

1. Joe is in his thirties, and _____*so is*_____ Sam.

2. Joe will be forty in a few years, and Sam _____.

3. Joe isn't a music teacher, and Sam _____.

4. Joe is married, and Sam _____.

5. Joe isn't single, and _____ Sam.

6. Joe has two daughters, and _____ Sam.

7. Joe doesn't have any sons, and Sam _____.

8. Joe lives in a big city, and Sam _____.

9. Joe doesn't eat meat, and _____ Sam.

10. Joe has been to London, and _____ Sam.

11. Joe is studying a foreign language, and Sam _____.

QUIZ 7 Connecting Ideas with *Because* (Chart 8-6)

Directions: Combine each pair of sentences in two different orders. Use ***because***. Punctuate the new sentences carefully.

Example: Andy is going to go shopping. His favorite store is having a sale.
> Andy is going to go shopping because his favorite store is having a sale.
> Because Andy's favorite store is having a sale, he is going to go shopping.

1. Sue really likes sports shoes. She spends a lot of money on shoes.

2. Cindy was driving too fast. She had an accident.

3. We went to the beach. It was a beautiful day.

4. I need to get new jeans. My old jeans have holes in them.

5. My car is making strange noises. I feel uncomfortable driving.

QUIZ 8 Connecting Ideas: *So* and *Because* (Charts 8-3 and 8-6)

Directions: Complete the sentences with ***so*** or ***because***.

Example: I wore my sunglasses _____*because*_____ the sun was shining.

1. Yesterday I had the flu, _____ I didn't go to work.

2. _____ my mother's birthday is tomorrow, I am going to buy her some flowers.

3. Harry is nervous about meeting his girlfriend's parents _____ he wants to marry her.

(continued on next page)

4. Dave plays in a jazz band, _____ he is often busy on Saturday nights.

5. Campers in Yellowstone Park have to be careful _____ there may be bears nearby.

6. My feet hurt, _____ I took my shoes off.

7. _____ the print is so small, I can't read the instructions on this package.

8. The musicians have to practice a lot _____ their concert is next weekend.

9. Tammy Riverton is funny and clever, _____ many people like her late-night TV show.

10. _____ Mike usually doesn't sleep well at night, he often feels tired during the day.

QUIZ 9 Connecting Ideas: *Even Though / Although* and *Because* (Charts 8-6 and 8-7)

A. *Directions:* Choose the correct sentence in each pair.

Example: (a.) Because the page is torn, I can't read it.
   b. Even though the page is torn, I can't read it.

1. a. The skaters practice every day although they hope to skate in the Olympics.
   b. The skaters practice every day because they hope to skate in the Olympics.

2. a. The store is open at night even though there aren't many customers.
   b. The store is open at night because there aren't many customers.

3. a. Although we try to save money, we always seem to spend more than we have.
   b. Because we try to save money, we always seem to spend more than we have.

4. a. Even though Tony twisted his ankle, he can't run in the race.
   b. Because Tony twisted his ankle, he can't run in the race.

5. a. Although the police could smell gas, they couldn't find a gas leak.
   b. Because the police could smell gas, they couldn't find a gas leak.

(continued on next page)

B. Directions: Choose the best completion for each sentence.

Example: The students stayed indoors at school because _____

    a. they wanted to play soccer.

    ⓑ. there were storm clouds in the area.

1. Because our car needed gas, _____

   a. we stopped at the gas station.

   b. we didn't stop for gas before we left town.

2. Nora didn't go to the dentist even though _____

   a. she had a toothache.

   b. she didn't have a toothache.

3. The kitchen smells delicious because _____

   a. we burned a pizza.

   b. we have been baking cookies.

4. Although Alan _____, he has difficulty getting up on time.

   a. sets two alarms on his phone

   b. stays up late

5. Even though my teacher told us about the test, _____

   a. I studied hard.

   b. I didn't study at all.

QUIZ 10 Punctuating Adverb Clauses (Charts 8-6 and 8-7)

Directions: Add commas as necessary.

Example: Even though I like eggplant**,** I rarely eat it.

1. Erica can't figure out this puzzle even though she did it once before.
2. Because the students felt the building shake they got under their desks.
3. Although there was a lot of traffic we got home on time.
4. Carlos can't eat peanut butter because he is allergic to peanuts.
5. Alice's Wi-Fi is slow even though she just got a new phone.
6. Because my parents were celebrating their 25th anniversary they had a big party.
7. A lot of people are working even though it's a holiday.
8. Even though it was raining we stood in line to get tickets for the concert.
9. It's difficult for me to do crossword puzzles even though I love word games.
10. Although this computer is new it starts up slowly.

QUIZ 11 Chapter Review

Directions: Correct the errors. More than one answer may be correct.

Example: Clothing is getting more expensive, and ~~neither~~ ^so is food.

1. I enjoy science my favorite subjects are physics math and chemistry.
2. Julia doesn't participate in sports. Either her friends.
3. Our baseball team lost the game. Because not enough players showed up.
4. My phone isn't working well and Jack's doesn't either.
5. I wore a hat and sunglasses at the beach. So is my sister.
6. My mother is Australian, my father is Brazilian.
7. Even though you're upset now, but you'll understand our decision in a few days.
8. Because our parents both work, so my brothers and I sometimes cook dinner.
9. I have never been to Hawaii, and my husband too.
10. The photographs turned out wonderfully, but the video isn't.

CHAPTER 8 – TEST 1

Part A *Directions:* Add commas, periods, and capital letters as necessary. Don't change any words or the order of the words.

SITUATION: *An Afternoon at the Beach*

1. Elena decided the weather was too nice to stay at home so she packed a picnic lunch and drove to the beach

2. even though it was crowded she found a place to sit she spread out her blanket and opened her lunch box

3. inside was a sandwich potato chips and an apple because she was still full from breakfast she ate only a little and saved the rest for later

4. she took out a book and opened it minutes later she was asleep and she woke up just as the sun was going down

Part B *Directions:* Complete the sentences with *and, but, or, so, because,* or *even though*.

SITUATION: *Bryan and Cathy's Wedding*

1. Bryan and Cathy are getting married soon, _____ they are very busy.

2. _____ there is a lot to prepare for the wedding, Bryan doesn't seem stressed.

3. They invited many people, _____ not everyone can come.

4. _____ Cathy's aunt has a large home, she has invited the whole family for a wedding party.

5. We can buy their gift online today, _____ we can go shopping next Saturday.

6. Cathy will wear a beautiful dress, _____ Bryan will wear a three-piece suit.

7. _____ the wedding is next month, Cathy hasn't bought a wedding dress yet.

8. Who's going to sing at the wedding, Mary _____ someone else?

9. They love chocolate, _____ they'll have a chocolate wedding cake.

10. Bryan's boss won't be able to come to the wedding _____ he'll be away on business.

Connecting Ideas: Punctuation and Meaning   133

Part C *Directions:* Complete the sentences with **so, too, either,** or **neither** and the correct auxiliary verb.

SITUATION: *Students in My English Class*

1. Nini likes learning English, and _____ Kira.
2. Dao is from Vietnam, and Boon _____.
3. Johannes isn't from Asia, and Martin _____.
4. Jacob doesn't speak Chinese, and _____ Masaki.
5. Max speaks German, and _____ I.
6. Bashir speaks Russian, and Andre _____.
7. Shirley hasn't been in the United States long, and _____ Julien.
8. Erik will go to Disneyland during spring break, and Vincent _____.
9. Mona is from Saudi Arabia, and _____ Majed and Ibrahim.
10. The teacher enjoys this class, and the students _____.

Part D *Directions:* Correct the errors. More than one answer may be correct.

1. I study hard even my classes are very easy.
2. After the accident, my left arm hurt and too my right shoulder.
3. Blackberries, strawberries, blueberries. They all grow in our garden.
4. Because the parking fees were so high, and people didn't want to park there.
5. We were excited about the concert, but we got there early to get good seats.
6. Kate was hungry. So she ate a sandwich.
7. Georgie will drive to Seattle, and so does Ken.
8. Although I was tired, but I stayed up late and finished my homework.

CHAPTER 8 – TEST 2

Part A *Directions:* Add commas, periods, and capital letters as necessary. Don't change any words or the order of the words.

SITUATION: *Ron's Future*

1. Ron needs to decide if he is going to go to graduate school or if he is going to get a job

2. he will finish business school in a few months although he has enjoyed being a student he is looking forward to graduation

3. his parents want him to get a master's degree they have said they will pay for it so they think he should agree to stay in school

4. Ron appreciates their generosity but he also wants to be more independent at this time in his life he wants to start earning his own money

Part B *Directions:* Complete the sentences with **and, but, or, so, because,** or **even though.**

1. Do you want some eggs? I can make them fried _____ boiled.

2. _____ the neighbors are noisy in the morning, Joyce wakes up early.

3. Sometimes my parents shop online, _____ they prefer to go to the mall.

4. I fell asleep at the train station, _____ I missed my train.

5. _____ Peter read the report several times, he still couldn't remember the important details.

6. Mrs. Davis took some time off from work _____ her son was in the hospital.

7. We went to several stores, _____ we couldn't find a birthday gift for Nathan. It's not easy to shop for him.

8. _____ Melissa slept eight hours, she still felt tired in the morning.

9. Do Martin and Kay live on First Street, _____ is it Fifth Street?

10. The owner wanted too much money for his truck, _____ we decided not to buy it.

Part C *Directions:* Complete the sentences with **so, too, either,** or **neither** and the correct auxiliary verb.

SITUATION: *A Day at the Zoo*

1. I am at the zoo, and _____ my daughter.
2. The penguins seem happy, and the monkeys _____.
3. The zebras live in a grassy area, and _____ the giraffes.
4. A big elephant is having a bath, and a small one _____.
5. The bears are eating, and _____ the gorillas.
6. The lions will get food later, and _____ the tigers.
7. Elephants don't eat meat, and gorillas _____.
8. We ate lunch at the Zoo Café, and _____ many other people.
9. My daughter likes the birds, and I _____.
10. She doesn't like the snakes, and _____ I.

Part D *Directions:* Correct the errors. More than one answer may be correct.

1. People couldn't describe the accident. Because it happened so quickly.
2. Even Nadia is a new student, but she has made many friends.
3. So a storm was approaching the sailors decided to go into shore.
4. You can pay either by cash and check. Which do you prefer?
5. Maria didn't understand the lecture. Neither I did too.
6. My mom grew up in a small town because she doesn't like to drive in the city.
7. I won't be home tonight and so won't my wife.
8. Harry was hungry at 9:00 A.M. even though he didn't eat breakfast before school.

CHAPTER 9 Comparisons

QUIZ 1 Comparatives (Chart 9-1)

Directions: Complete the sentences with the correct comparative form (*-er / more*) of the adjectives in parentheses.

Example: The Taj Mahal is (*old*) __older__ than the Blue Mosque.

1. The Sahara Desert is (*large*) _____ than the Kalahari Desert.
2. Brown rice is (*good*) _____ than white rice.
3. Which is (*dangerous*) _____: ice climbing or motorcycle racing?
4. A basketball is (*heavy*) _____ than a balloon.
5. An essay is (*long*) _____ than a paragraph.
6. Small grocery stores are usually (*expensive*) _____ than supermarkets.
7. Paying bills online can be (*easy*) _____ than writing checks.
8. Taking the subway is (*quick*) _____ than taking a bus.
9. I find action movies (*exciting*) _____ than romantic movies.
10. Riding on a motorbike with a helmet is (*safe*) _____ than riding without one.

QUIZ 2 Comparatives (Chart 9-1)

Directions: Make comparison sentences with *-er / more* and the adjectives in parentheses.

Example: a turtle \ a rabbit (*fast*)
 _____A rabbit is faster than a turtle._____

1. summer \ winter (*warm*)
 _____

2. a tiger \ a cat (*big*)
 _____

3. a day at the beach \ a day at work (*relaxing*)
 _____

4. a comedy \ a drama (*funny*)
 _____

(continued on next page)

5. snow \ rain (*cold*)

_____

6. strawberries \ lemons (*sweet*)

_____

7. a car \ a bicycle (*expensive*)

_____

8. a rock \ a flower (*hard*)

_____

9. taking tests \ doing homework (*stressful*)

_____

10. a year \ a month (*long*)

_____

QUIZ 3 Superlatives   (Chart 9-2)

Directions: Complete the sentences with the correct superlative form (*-est / most*) of the adjectives in parentheses.

Example: DePaul's Mercantile is (*busy*) _____the busiest_____ store in town.

1. My brother is (*friendly*) _____ person in my family.

2. These are (*big*) _____ tomatoes I've ever seen.

3. Wellington, the capital of New Zealand, is (*windy*) _____ city in the world.

4. Mary is (*good*) _____ music teacher I know.

5. Angelina's Café is (*expensive*) _____ restaurant in town.

6. My sister (*smart*) _____ person in her class.

7. The blue whale is (*large*) _____ animal on earth.

8. Los Angeles has (*bad*) _____ traffic in the USA.

9. A passport is (*important*) _____ document for international travel.

10. These running shoes are (*comfortable*) _____ shoes I have.

QUIZ 4 Completing Comparatives (Chart 9-3)

Directions: Complete the sentences. Use pronouns in the completions. The first one is done for you.

SITUATION: *The Smith Family*

1. You don't know the Smiths well. I know them better than _____*you do*_____.

2. Dr. Gary Smith is 33 years old. His wife, Annie, is 30. I am 29. Gary and Annie are older than _____. I am younger than _____.

3. The Smiths have four children. We only have two children. Their family is bigger than _____.

4. The Smith's son, Jared, can count to 100 in Spanish. The other children in his class can only count to 20. Jared can count better than _____.

5. Annie doesn't enjoy cooking. Her husband loves to cook. His cooking is better than _____.

6. Gary can run a mile in 10 minutes. Annie can run a mile in eight minutes. Annie can run faster than _____.

7. We bought a two-bedroom house. The Smiths bought a four-bedroom house. The Smiths bought a larger house than _____.

8. Dr. Smith isn't an expensive dentist. The other dentists in the city are more expensive than _____.

9. The Smith's daughter, Beth, loves to swim. Her brothers and sisters don't like to swim that much. Beth likes swimming more than _____.

10. The Smiths have a dog. Gary doesn't have time to walk the dog, so Annie usually does it. Annie walks the dog more often than _____.

QUIZ 5 Completing Superlatives (Chart 9-3)

A. *Directions:* Complete the sentences with the superlative form of the adjectives in parentheses and ***in, of,*** or ***ever.***

Example: The kitchen is (*warm*) ___the warmest___ room ___in___ our house.

1. There are several talented singers in our choir, but Marcos Rivera has (*beautiful*) _____ voice _____ all.

2. Xander's Taverna is (*good*) _____ restaurant _____ Marysville.

3. This rental car is the (*fast*) _____ car I've _____ driven.

4. Mr. Jones works (*hard*) _____ all the people in his office.

5. Dana's Fashions is (*busy*) _____ store _____ the shopping mall.

B. *Directions:* Complete the sentences with the superlative form of the words in ***italics***. Include the correct form of the noun in your answer.

Example: Teresa is a ***lazy girl***. She's one of ___the laziest girls___ in our class.

1. I know many ***nice people***, but my sister Julie is one of _____ I know.

2. Siberia is a ***cold place***. In fact, it's one of _____ in the world.

3. I had ***really bad day*** at work yesterday. It was one of _____ I've ever had.

4. Ann likes museums and has seen many ***beautiful paintings***. "Starry Night" by Van Gogh is one of _____ she has ever seen.

5. My hotel room had a very ***comfortable bed***. That was one of _____ I've ever slept on.

QUIZ 6 Comparatives vs. Superlatives (Charts 9-1 → 9-3)

Directions: Circle the correct completions.

Example: My brother is **(taller than)/ the tallest** my dad.

1. a. Jason eats a lot. He is always **hungrier than / the hungriest** I am.
   b. Last week he ate **bigger than / the biggest** sandwich I've ever seen!
2. a. My math teacher is **more helpful / the most helpful** teacher in our school.
   b. For some students, math is **more difficult than / the most difficult** English.
3. a. The meat in this store is expensive, and steak is **more expensive than / the most expensive** of all.
   b. I usually buy chicken. It is usually **cheaper than / the cheapest** beef.
4. a. Chris and Marta's house is **older than / the oldest** our house.
   b. It's also **more beautiful than / the most beautiful** house in the neighborhood.
5. a. Last winter, we had **colder than / the coldest** weather we've ever had.
   b. This year, it is **wetter than / the wettest** last year.

QUIZ 7 Comparisons with Adverbs (Chart 9-4)

Directions: Complete the sentences with the correct comparative or superlative form of the adverbs in parentheses.

Example: He lives much (*far*) _____farther_____ from school than I do.

1. Maxine always washes the dishes (*fast*) _____ than Artie.
2. I study (*hard*) _____ than my brother.
3. My grandfather drives (*carefully*) _____ than my grandmother.
4. Dr. Woodward arrives at work (*early*) _____ of all the doctors in the clinic.
5. Taiga writes Japanese characters (*well*) _____ than his younger sister.
6. Josh finished his exam (*quickly*) _____ than Sarah.
7. Everyone in my dad's company works hard, but my dad works (*hard*) _____ of all.
8. Miko came to school (*early*) _____ than his classmates.
9. In the video game competition, Nick did (*well*) _____ of all the people on his team.
10. Our English teacher speaks (*slowly*) _____ than our other teachers.

QUIZ 8 Farther and Further (Chart 9-4)

Directions: Complete the sentences with **farther** or **further**. Use both if possible.

Example: The bank is ___farther / further___ from here than the post office.

1. Don ran five kilometers, but he couldn't run _____. He was too tired.

2. Please think about this _____ before you make a decision.

3. My grandparents like their new retirement community, but now they live _____ from town than they did before.

4. If you need more information, please contact my lawyer. I can't discuss it _____.

5. We hiked all the way to the lake! We went _____ than we had planned.

6. Hong Kong is _____ from Beijing than from Shanghai.

7. Chelsea is interested in renewable energy. She wants to research it _____.

8. Since my surgery, I've begun taking walks. Every day, I try to walk a little _____.

9. The town council needs to study the problem _____ so they can find the best solution.

10. The bus station is _____ from our house than the train station.

QUIZ 9 Repeating a Comparative (Chart 9-5)

Directions: Complete the answers by repeating the comparatives in *italics*.

Example: Charles is so **tall**. He has really grown! He is getting ___taller and taller___.

1. The weather has been **hot** lately. According to this week's weather forecast, it will get _____.

2. My teacher doesn't explain geometry problems clearly. I feel **confused**. When I ask my teacher to explain things again, I feel _____.

3. While we were hiking, it started to rain **hard**. It rained _____, so we decided to quit and go home.

4. When Jerry starts to laugh, his face turns **red**. As he continues to laugh, his face gets _____.

5. Each term, my classes seem more **difficult**. As they become _____, I'm studying harder and learning more.

(continued on next page)

6. Oscar was not *relaxed* with his children when they were younger. As they get older, he's becoming _____.

7. I lost five pounds, so I was very *happy*. As I continued to lose weight, I felt _____.

8. Medical insurance is very *expensive*. It seems that every year, it gets _____.

9. Snowboarding competitions are *exciting*. In fact, snowboarding competitions seem to get _____ every year.

10. This story is so *good*. It gets _____ with every chapter.

QUIZ 10 Using Double Comparatives (Chart 9-5)

A. Directions: Complete the sentences with double comparatives (*the more / -er … the more / -er*) and the words in *italics*.

Example: I love *cold* drinks. They are *refreshing*.

The ____colder____ a drink is, ____the more refreshing____ it is.

1. I love *thick* steaks. They are *good*.

   _____ a steak is, _____ it is.

2. Mike talks *fast*. It's *difficult* to understand him.

   _____ Mike talks, _____ it is to understand him.

3. The weather is *cold*. I feel *bad*.

   _____ the weather is, _____ I feel.

4. Lucy studies *hard*. She scores *high* on exams.

   _____ Lucy studies, _____ she scores on exams.

5. My parents took me to a very *nice* restaurant on my birthday. It was very *expensive*.

   _____ a restaurant is, _____ it is.

(continued on next page)

B. Directions: Combine each pair of sentences. Use double comparatives (*the more / -er ... the more / -er*) and the words in *italics*.

Example: The couple argued **loudly**. They became **upset**.

<u>The more loudly they argued, the more upset they became.</u>

1. It got **dark**. The children were **afraid**.
   _____

2. I **exercised**. I felt **energetic**.
   _____

3. The story was **funny**. The children laughed **hard**.
   _____

4. Bob arrives at work **late**. His boss gets **impatient**.
   _____

5. Simon works **fast**. His boss feels **happy**.
   _____

QUIZ 11 Modifying Comparatives   (Chart 9-6)

Directions: Complete the sentences with ***very, much, a lot,*** or ***far***. More than one answer may be correct.

Example: Channel 8 has _____<u>a lot</u>_____ more international news than other TV stations.

1. This fish doesn't taste _____ good. You don't have to eat it.

2. My mother is _____ more active than my father.

3. It's _____ cold outside. Be sure to wear your warm coat.

4. Barbara has a sunny personality. She is _____ happier than most people I know.

5. For me, math is _____ more difficult than a foreign language.

6. Len looks _____ tired. Is he getting enough sleep?

7. Tasha's _____ excited about school. She will be in kindergarten this year.

8. This watch is _____ nicer than I remembered. It's also _____ more expensive than I remembered.

9. Janice is _____ older than her husband.

QUIZ 12 Negative Comparisons (Chart 9-7)

Directions: Choose the sentence in each pair that is closest in meaning to the given sentence.

Example: His bedroom has never been dirtier.

    (a.) His bedroom is dirty.   b. His bedroom isn't dirty.

1. I've never seen a better movie.

    a. I liked the movie very much.   b. I didn't like the movie much.

2. Rob has never spoken Japanese.

    a. Rob speaks Japanese.   b. Rob doesn't speak Japanese.

3. I've never read a more terrible book.

    a. I liked the book a lot.   b. I didn't like the book at all.

4. Jim has never made spicier curry.

    a. The curry is very spicy.   b. The curry isn't spicy.

5. Francesca has never been strong.

    a. Francesca is strong.   b. Francesca isn't strong.

6. I've never met a friendlier dog.

    a. The dog is friendly.   b. The dog isn't friendly.

7. Teddy has never had a worse cold.

    a. Teddy has a cold.   b. Teddy doesn't have a cold.

8. Kate has never been happier.

    a. Kate is happy.   b. Kate isn't happy.

9. I've never liked chicken noodle soup.

    a. I like chicken noodle soup.   b. I don't like chicken noodle soup.

10. My family has never been prouder of me.

    a. My family is very proud of me.   b. My family isn't proud of me.

QUIZ 13　Comparisons with As … As　(Chart 9-8)

A. Directions: Compare the temperatures of the cities. Use **just as … as**, **almost as … as**, or **not as … as**. The first one is done for you.

Yesterday's weather:	Vienna	68°F / 20°C	Hong Kong	75°F / 24°C
	Vancouver	70°F / 21°C	Athens	90°F / 32°C
	Paris	70°F / 21°C	Dubai	104°F / 40°C

1. Paris was ____not as warm as____ Athens.

2. Athens was _____ Dubai.

3. Vienna was _____ Paris and Vancouver.

4. Vancouver was _____ Paris.

5. Vienna and Paris were _____ Athens.

6. Hong Kong was _____ Athens and Dubai.

B. Directions: Compare the ages of the people. The first one is done for you.

Jim	27	Max	40
Susan	28	Paulo	43
Lacey	28	Maria	45

1. Susan is as old as ____Lacey____ .

2. Paulo is almost as old as _____ .

3. Max is not quite as old as _____ .

4. Lacey is just as old as _____ .

5. Jim is nearly as old as _____ .

6. Paulo is not as old as _____ .

QUIZ 14 Comparisons with *As ... As* (Chart 9-8)

Directions: Using the given words, write sentences with *just as ... as* or *not as ... as*. Give your own opinions.

Example: a sunrise / a sunset (*beautiful*)

    A sunrise is just as beautiful as a sunset .

1. light chocolate / dark chocolate (*delicious*)

   _____ .

2. spring / fall (*colorful*)

   _____ .

3. sending email / text messaging (*easy*)

   _____ .

4. a hard pillow / a soft pillow (*comfortable*)

   _____ .

5. a poisonous snake / a non-poisonous snake (*scary*)

   _____ .

6. eating / sleeping (*important*)

   _____ .

7. history / biology (*difficult*)

   _____ .

8. a bicycle / a motorbike (*fast*)

   _____ .

9. soccer / basketball (*popular*)

   _____ .

10. the Winter Olympics / the Summer Olympics (*exciting*)

    _____ .

QUIZ 15 Less ... Than and Not As ... As (Chart 9-9)

Directions: Circle the correct completions. In some cases, both answers are correct.

Example: Gorillas are (less intelligent than)/(not as intelligent as) chimpanzees.

1. Green peppers are   **less sweet than / not as sweet as**   carrots.
2. A shower is   **less relaxing than / not as relaxing as**   a bath.
3. High heels are   **less comfortable than / not as comfortable as**   slippers.
4. Sweets are   **less healthy than / not as healthy as**   vegetables.
5. Riding a bike is usually   **less fast than / not as fast as**   driving.
6. This test is   **less difficult than / not as difficult as**   the last one.
7. The blue pillow is   **less soft than / not as soft as**   the green pillow.
8. The fish from the grocery store is   **less expensive than / not as expensive as**   the fish from the fish market.
9. Country air is   **less fresh than / not as fresh as**   city air.
10. Arithmetic is   **less hard than / not as hard as**   calculus.

QUIZ 16 Comparatives with Nouns, Adjectives, and Adverbs (Charts 9-1, 9-4, and 9-10)

Directions: Make comparisons with the words in parentheses. Add **more / -er** as needed.

Example: I collect dolls. I have (*dolls*) _____more dolls_____ than my sisters have.

1. A top-loading washing machine uses (*water*) _____ than a front-loading machine.
2. Shelley runs (*fast*) _____ than the rest of the track team.
3. You have ten minutes left. Is that enough, or do you need (*time*) _____?
4. Our literature class this year is (*difficult*) _____ than our literature class last year.
5. The weather is (*sunny*) _____ today than yesterday.
6. This used car has (*problems*) _____ than I expected.
7. The last math test was pretty hard. I hope the next one is (*easy*) _____.
8. The grass outside has turned brown. We need (*rain*) _____.
9. I like cold weather (*well*) _____ than my husband does.
10. The students are working (*quietly*) _____ today than they were yesterday.

QUIZ 17 The Same, Similar, Different, Like, and Alike (Chart 9-11)

A. Directions: Complete the sentences with **as, to, from,** or **Ø**.

Example: Sometimes Spiro looks like ___Ø___ he's not listening, but actually he's just concentrating.

1. My three-year-old son thinks that donkeys look similar _____ zebras. He says they look alike _____ because of their ears.

2. My new house is similar _____ my old house, but it's more updated. Fortunately, the washer and dryer are different _____ my old ones. They're much newer.

3. A: Is your hometown much different _____ this city?

   B: The size of this town is the same _____ my hometown, but the buildings look very different _____ .

4. A panda is like _____ a bear, but they are not the same _____ . Bears don't have thumbs, but pandas do.

B. Directions: Compare the size of the clouds. Use **the same (as), similar (to), different (from), like,** or **alike**. The first one is done for you.

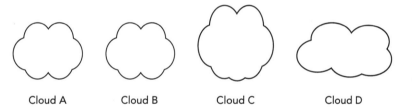

Cloud A   Cloud B   Cloud C   Cloud D

1. Cloud A is ___the same as___ Cloud B.

2. Cloud A and Cloud B are _____ .

3. Cloud A and Cloud D are _____ .

4. Cloud C is _____ Cloud D.

5. All of the clouds are _____ each other.

6. All of the clouds are _____ .

QUIZ 18 The Same, Similar, Different, Like, and Alike (Chart 9-11)

Directions: Complete the sentences, using *the same, similar, different, like,* or *alike*. More than one answer may be correct.

Example: A chipmunk looks ___similar___ to a squirrel.

1. Dick likes comedy, and Sharon likes drama. Their movie preferences are very _____.

2. Franz wants to marry someone who thinks _____ him and agrees with him on important things.

3. Tom and Will are cousins, but people think they are brothers. They have _____ eye color and hair color, and they talk and walk _____.

4. "They're," "their," and "there" all have _____ pronunciation.

5. A: How is the Pacific Ocean _____ from the Atlantic Ocean?

   B: The Pacific Ocean is larger and deeper than the Atlantic Ocean. It's also not as salty.

6. I'm sorry, but we can't accept your credit card. The signature on the card and the one on the receipt are not _____.

7. "Teacher, Alejandro and I have exactly the same answers on the test, but we got _____ grades. Why?"

8. Since many big cities have _____ traffic and transportation problems, city leaders can learn about solutions from each other.

9. My twin nephews look exactly _____, so I can't tell them apart.

QUIZ 19 Chapter Review

Directions: Correct the errors.

Example: This coffee should be ~~more~~ warmer. Could you heat it up, please?

1. The most friendliest person in our class is Julie.
2. The food at the school cafeteria is less good than a lunch from home.
3. The movie was very funny than we expected. We laughed the whole time.
4. Anna's dog is ugliest dog I've ever seen.
5. Grandpa's behavior is embarrassing. The older he gets, the loudly he talks.
6. I have a same bag as you. Where did you buy yours?
7. Venice is one of the most interesting city I've ever visited.
8. For Jon, the game of chess is alike an interesting math puzzle.
9. As the horse got tired, he began walking slower and more slow.
10. The driving test was more hard than I expected.

CHAPTER 9 – TEST 1

Part A *Directions:* Complete the sentences with the comparative or superlative form of the adjectives in parentheses.

1. Did you know that cold water is (*heavy*) _____ warm water?

2. Which is (*small*) _____ : the moon or the earth?

3. In our town, December is very cold and windy. Sometimes, December is (*cold*) _____ January. But the summers are hot, and August is (*hot*) _____ of all the months.

4. Which movies do you think are (*bad*) _____ for children: movies with violence or movies with bad language?

5. Tokyo is (*large*) _____ city in Japan. It is also one of (*crowded*) _____ cities in the world.

6. It's (*easy*) _____ to drive on State Street _____ on Park Avenue because State Street has less traffic.

7. Did you know that Greenland is (*big*) _____ island in the world?

8. The cheesecake was wonderful. It was one of (*delicious*) _____ cheesecakes I've ever eaten.

Part B *Directions:* Write sentences comparing the size of the circles. Use the words in parentheses and *big* or *small* where appropriate. More than one answer may be correct.

Circle A   Circle B   Circle C   Circle D   Circle E

1. (*as ... as*) _____.

2. (*not as ... as*) _____.

3. (*different*) _____.

4. (*the ... -est*) _____.

5. (*almost as ... as*) _____.

Part C *Directions:* Complete the sentences, using *in*, *of*, *as*, *to*, *from*, or *Ø*.

1. That was the most difficult test _____ all the tests we've taken.

2. White bread is not as healthy _____ whole wheat bread.

3. Who is the funniest student _____ your class?

4. I think Professor Brown's teaching methods are very different _____ Professor Green's.

5. Mrs. Thompson thinks it's cute to dress her twins alike _____.

6. A pond is similar _____ a lake, but it is smaller.

7. The food at Tim's Diner is not as good _____ the food at Sunny's restaurant.

8. Henry is a fast worker, and he is the youngest _____ all the employees.

Part D *Directions:* Circle the correct completions.

1. Olivia and Robert have a new baby and aren't sleeping much. They have never felt _____.
   a. more tired   b. more tired than   c. most tired

2. That's an interesting idea. I need to research it _____ before I make a decision.
   a. farther   b. the furthest   c. further

3. Mark has a loud voice, but he talks _____ when his children are asleep.
   a. quieter   b. more quietly   c. more quietly than

4. I'm not rich. Many people have more money _____.
   a. than I am   b. than I do   c. than my

5. Johnny was adding air to a balloon. The balloon got _____. Then it popped!
   a. more and more big   b. more bigger   c. bigger and bigger

6. I'm not very hungry. Carlos is _____ than I am.
   a. much hungrier   b. far hungry   c. very hungrier

7. Victor has a 30-minute walk to school. Of all the kids in our class, he lives _____ away.
   a. farther   b. further   c. the furthest

Part E *Directions:* Correct the errors.

1. Let's buy this chair. It's less expensive from that one.
2. My brother is shorter than mine.
3. Linda is in as same German class as I am.
4. What was a happyiest day in your life?
5. The kids yelled loudly and more loudly in the park.
6. If you need farther help, please ask.
7. My homework isn't as difficult to yours.
8. Erin's stomachache got worser and worser as the day went on.
9. That was one of the best book I have ever read.
10. Those students are the most smartest kids in the class.

CHAPTER 9 – TEST 2

Part A *Directions:* Complete the sentences with the comparative or superlative form of the adjectives in parentheses.

1. In Thai cooking, green curry usually tastes (*spicy*) _____ yellow curry.

2. On our trip to the Grand Canyon, we saw some of (*lovely*) _____ scenery we have ever seen.

3. Is a kilometer (*long*) _____ or (*short*) _____ a mile?

4. Some artists prefer to paint with watercolors, but I think oils are (*interesting*) _____ watercolors.

5. The Caspian Sea is (*big*) _____ lake in the world.

6. Our essay topic for next week is (*bad*) "_____ person in history." I wish the teacher had given us (*famous*) "_____ person in history."

7. Highway 93 is one of (*dangerous*) _____ roads in Montana.

8. Nancy always stops to say hello. She is (*friendly*) _____ our neighbors.

Part B *Directions:* Write sentences comparing the height of the trees. Use the words in parentheses and **tall** or **short** where appropriate. More than one answer may be correct.

Tree A   Tree B   Tree C   Tree D   Tree E

1. (*similar*) _____.

2. (*just as ... as*) _____.

3. (*not quite as ... as*) _____.

4. (*-er ... than*) _____.

5. (*the same*) _____.

Part C *Directions:* Complete the sentences with *in, of, as, to, from,* or *Ø*.

1. A maple tree does not grow as tall __as__ a redwood tree.
2. What is the coldest place _____ the world?
3. The public transportation system in this city is very different _____ the system in my country. My country has more subways and buses.
4. How are a duck and a goose alike _____?
5. Kuala Lumpur was definitely the hottest _____ all the cities I've traveled to.
6. My cousin's hairstyle is very similar _____ mine.
7. Michelle is the most experienced manager _____ the company.
8. Mary is the funniest _____ all my sisters.

Part D *Directions:* Circle the correct completions.

1. How much ____ do we have to drive?
   a. farther than   b. farther   c. the furthest
2. My wife drinks ____ than I do.
   a. more coffee   b. coffee   c. the most coffee
3. John really likes his new job. He has never looked ____.
   a. happiest   b. more happy   c. happier
4. It's a hot day, so I'm really thirsty. The hotter it is, ____.
   a. the thirstiest   b. the thirstier I am   c. I'm thirstier
5. Teddy is 20 years old, but his sister is only eight. He is ____ than his sister.
   a. very older   b. much more older   c. far older
6. This book is ____ the last book I read.
   a. not so interesting as   b. less interesting as   c. not as interesting as
7. Let's discuss this idea ____ when we get home.
   a. farther   b. further   c. farther than

Part E *Directions:* Correct the errors.

1. These peas are delicious. Fresh peas taste so much better as frozen peas.
2. The flu can be a dangerous illness. It's very dangerous to have the flu than a cold.
3. Pluto is the smallest the nine planets.
4. The clouds look dark. Let's hope it's not as rainy today than it was yesterday.
5. Who has a more good life: a married person or a single person?
6. I'm a morning person. I have most energy in the morning.
7. As we skied down the mountain, we went more and more fast.
8. The ostrich is a largest bird, but the elephant is the largest land animals.
9. Titanium is one of the strongest metal in the world.
10. Please talk more quiet in the library.

CHAPTER 10 The Passive

QUIZ 1 Active or Passive (Charts 10-1 and 10-2)

Directions: Decide if the sentence is active or passive. The first one is done for you.

1. William brought a big gift to the wedding. (active) passive
2. My hair was cut by the barber. active passive
3. The teacher will correct the errors on the test. active passive
4. The package was delivered by the mail carrier. active passive
5. Emma was invited to the wedding. active passive
6. The website is used by millions of people around the world. active passive
7. The flowers were eaten by a deer. active passive
8. Brandon posted a picture on his web page. active passive
9. The movie was written by the director. active passive
10. The food was prepared by a famous chef. active passive
11. The cat has scratched the furniture. active passive

QUIZ 2 Forming the Passive (Charts 10-1 and 10-2)

Directions: Change the active verbs to passive by adding the correct forms of the verb **be**.

SITUATION: *Writing a Report*

Example: The boss reads the report. The report _____*is*_____ read by the boss.

1. Amy writes the report.

    The report _____ written by Amy.

2. The managers write the reports.

    The reports _____ written by the managers.

3. Amy wrote the report.

    The report _____ written by Amy.

4. The managers wrote the reports.

    The reports _____ written by the managers.

(continued on next page)

5. Amy has written the report.

   The report _____ written by Amy.

6. The managers have written the reports.

   The reports _____ written by the managers.

7. Amy will write the report.

   The report _____ written by Amy.

8. The managers will write the reports.

   The reports _____ written by the managers.

9. Amy is going to write the report.

   The report _____ written by Amy.

10. The managers are going to write the reports.

    The reports _____ written by the managers.

QUIZ 3 Forming the Passive   (Charts 10-1 and 10-2)

Directions: Change the active verbs to passive. Do not change the verb tense.

Example: The doctor examined the patient.

   The patient ____*was examined*____ by the doctor.

1. Rob drives Martha to school every day.

   Martha _____ to school by Rob every day.

2. The baseball player hit a homerun.

   A homerun _____ by the baseball player.

3. The builder will fix our roof.

   Our roof _____ by the builder.

4. The students clean the classrooms.

   The classrooms _____ by the students.

5. The accountant has checked our tax forms.

   Our tax forms _____ by the accountant.

6. Mr. Fernandez wrote two books about train travel.

   Two books about train travel _____ by Mr. Fernandez.

(continued on next page)

7. Lucien ate the last piece of meat.

The last piece of meat _____ by Lucien.

8. Everyone enjoyed the parade.

The parade _____ by everyone.

9. Our gift is going to surprise you.

You _____ by our gift.

10. My boss signed the paperwork.

The paperwork _____ by my boss.

QUIZ 4 Understanding the Passive (Charts 10-1 and 10-2)

A. *Directions*: Choose the sentence that has the same meaning as the given sentence.

Example: The children were given some chocolate by their grandmother.

    a. The children gave their grandmother some chocolate.

    (b.) The grandmother gave the children some chocolate.

1. You will be given the information later.

   a. You will give someone the information.

   b. Someone will give you the information.

2. My sister was offered a new job by Mr. Crosby.

   a. My sister got a new job.

   b. Mr. Crosby got a new job.

3. The dog found a bone by the fence.

   a. The bone was found by the dog.

   b. The dog was found by the fence.

4. The story has been read to the children several times.

   a. The children have read the story.

   b. Someone has read the story to the children.

5. Chef Daniel cooks the guests' meals by himself on Monday nights.

   a. The meals are cooked by Chef Daniel.

   b. Someone else cooks the guests' meals.

(continued on next page)

B. *Directions*: Change the sentences from passive to active. Do not change the verb tense.

Example: The email was sent by my father.

My father sent the email.

1. Ice cream is enjoyed by children.

2. The book was discussed by the students.

3. The café's name has been changed by the new owners.

4. The game is going to be won by our team.

5. An award will be given to Mr. Reed by the city.

QUIZ 5 Active vs. Passive (Charts 10-1 and 10-2)

A. *Directions:* Change the statements and questions from active to passive.

SITUATION: *At the Hospital*

Example: a. The hospital will open a clinic.

A clinic will be opened by the hospital.

b. Will the hospital open a clinic?

Will a clinic be opened by the hospital?

1. a. Volunteers greet hospital visitors.

b. Do volunteers greet hospital visitors?

2. a. Patients speak many different languages.

b. Do patients speak many different languages?

(continued on next page)

3. a. The nurse has checked Adam's blood pressure.

   _____

   b. Has the nurse checked Adam's blood pressure?

   _____

4. a. The doctors discussed a new medication.

   _____

   b. Did the doctors discuss a new medication?

   _____

5. a. The hospital gives excellent care.

   _____

   b. Does the hospital give excellent care?

   _____

QUIZ 6 Progressive Forms of the Passive   (Chart 10-3)

A. Directions: Complete the passive sentences. Add the subject and the correct form of **be**. The first one is done for you.

SITUATION: *Cleaning Day*

1. My dad is mopping the kitchen floor.   _The kitchen floor is being_ mopped by my dad.
2. The kids are dusting the furniture.   _____ dusted by the kids.
3. My mom is helping me.   _____ helped by my mom.
4. Paul was vacuuming the carpets.   _____ vacuumed by Paul.
5. Gary was helping Paul.   _____ helped by Gary.
6. Nan was washing the windows.   _____ washed by Nan.

B. Directions: Change each sentence from passive to active. The first one is done for you.

SITUATION: *Holiday Preparations*

1. The house was being decorated by Rosa and Sam.

   Rosa and Sam were decorating the house.

2. Gifts were being bought by Mr. Lim.

   _____

3. Special cakes are being baked by my grandmother.

   _____

(continued on next page)

4. The whole family is being invited over by Aunt Lily and Uncle Ted.

5. A big dinner is being prepared by Aunt Lily.

6. The table is being set by Teresa and Louise.

QUIZ 7 Transitive and Intransitive Verbs (Chart 10-4)

Directions: Check (✓) *transitive* if the verb takes an object. Check (✓) *intransitive* if it doesn't. The first one is done for you.

	transitive	intransitive
1. Andy swam in the ocean.		✓
2. Julia met Antonio at the library.		
3. The lawyer solved the problem.		
4. The bus driver drove down the wrong street.		
5. The packages will arrive two weeks late.		
6. Gina eats eggs for breakfast every morning.		
7. Dick and Susan never agree with each other.		
8. The fish died after a week in the fish tank.		
9. The children broke the window with a ball.		
10. Mr. Park invited his parents to the theater.		
11. A storm hit the coast last night.		

QUIZ 8 Transitive vs. Intransitive (Chart 10-4)

Directions: Change the sentences to passive if possible. If not possible, write *no change*.

Example: The wind blew my hat across the garden.
   <u>My hat was blown across the garden by the wind.</u>

1. Jonathan was in a serious car accident last night.

(continued on next page)

2. The plumber has finally fixed our sink.

3. Paulo is going to leave before sunrise.

4. We arrived at the wedding on time.

5. Our neighbors will sell their car next month.

6. Mr. LeBarre usually washes the dinner dishes.

7. The Tangs stayed at a friend's summer house last month.

8. I slept for ten hours last night.

9. Val returned the broken lamp to the store.

10. Ivan is waiting for the bus.

QUIZ 9 Using the *By*-Phrase   (Chart 10-5)

Directions: Change the sentences from active to passive. Use the ***by***-phrase only when necessary.

Examples: Someone left his jacket on the bus.

> *A jacket was left on the bus.*

Millions of people watched the Olympics on TV.

> *The Olympics were watched on TV by millions of people.*

1. Someone gave me this sweater.

2. Larry Page and Sergey Brin created Google.

(continued on next page)

3. People check out books at a library.

   _____

4. Has anyone written the report yet?

   _____

5. Picasso painted the picture.

   _____

6. Someone will paint these walls tomorrow.

   _____

7. The police stopped the speeding car.

   _____

8. When did people first use cell phones?

   _____

9. People speak French and English in Canada.

   _____

10. The referee has stopped the basketball game.

    _____

QUIZ 10 Passive Modals   (Chart 10-6)

Directions: Complete the sentences by changing the active modals to passive.

Example: You shouldn't overcook vegetables.

   Vegetables _____*shouldn't be overcooked*_____ .

1. You can contact me by text or email.

   I _____ by text or email.

2. Someone has to take Ali to soccer practice now.

   Ali _____ to soccer practice now.

3. You shouldn't eat those berries.

   Those berries _____ .

4. You ought to send that package by express mail.

   The package _____ by express mail.

(continued on next page)

5. We have to change the date for our wedding.

   The date for our wedding _____.

6. Someone should give these old clothes to charity.

   These old clothes _____ to charity.

7. If you leave the doors unlocked, someone could steal your car.

   If you leave the doors unlocked, your car _____.

8. You must tell your parents about the accident.

   Your parents _____ about the accident.

9. The home team might win the game.

   The game _____ by the home team.

10. I think we may reach an agreement today.

    I think an agreement _____ today.

QUIZ 11 Review: Active vs. Passive (Charts 10-1 → 10-6)

Directions: Circle the correct completions.

Example: My brother **has chosen /(has been chosen)** to speak at graduation.

1. Our electric bill **has to pay / has to be paid** today.
2. Your dog **has eaten / has been eaten** all of the hot dogs.
3. Shahin **called / was called** his uncle by mistake. He got the wrong number.
4. Look at our dirty windows! They **should wash / should be washed** as soon as possible.
5. Pierre's birthday party **will hold / will be held** at Celebration Park.
6. You **can contact / can be contacted** Dr. Jones at his office on Monday.
7. Our apartment **painted / was painted** last weekend.
8. The Smith's summer cabin **built / was built** by their great grandfather.
9. I **turned off / was turned off** your computer.
10. The community **must tell / must be told** about the plans for new housing.

QUIZ 12 Past Participles as Adjectives   (Chart 10-7)

Directions: Complete each sentence with an appropriate preposition. More than one answer may be correct.

SITUATION: *Two Students*

Example: Eduardo and Carlos are related ___to___ each other. They are cousins.

Eduardo doesn't like school. He is ...

1. bored _____ his classes.
2. disappointed _____ his teachers.
3. never finished _____ his homework.
4. not involved _____ class projects.
5. tired _____ studying.

Carlos loves being a student. He is ...

6. excited _____ his classes.
7. interested _____ learning.
8. always prepared _____ class.
9. devoted _____ his studies.
10. satisfied _____ his grades.

QUIZ 13 Past Participles as Adjectives   (Chart 10-7)

Directions: Complete the sentences with a form of *be* and the past participle of the verbs in the box. Note the prepositions that follow them.

bore	engage	interest	please	~~scare~~	worry
crowd	excite	marry	qualify	tire	

Example: Some children ___are scared___ of the dark. They don't like to go to sleep in a dark room.

1. a. Ibrahim _____ for the job. I'm sure they will hire him.

   b. I _____ about my new job. I start next week, and I'm looking forward to it!

2. a. When Elena's alarm didn't ring, she _____ about being late for work.

   b. Marcia _____ with her job. She needs to find more interesting work.

3. a. Jay and Mark _____ from working a 12-hour day yesterday.

   b. The bus _____ on the way home. They couldn't find a seat.

4. a. Jacob _____ with the results of his driving test. He passed!

   b. Now he _____ in cars. He wants to buy a good used car.

5. a. Angela is so happy. She _____ to Yuri. Their wedding is in October.

   b. Julie and Fred _____ . They got married two years ago.

QUIZ 14 Participial Adjectives: -ed vs. -ing (Chart 10-8)

Directions: Complete the sentences with the appropriate *-ed* or *-ing* forms of the verbs in *italics*.

Example: Adam doesn't understand the problem. It **confuses** him.

The problem is _____confusing_____.

1. Camila watched the news. It **surprised** her.

    a. Camila was _____.

    b. The news was _____.

2. Jamal tore his pants. It **embarrassed** him.

    a. It was an _____ situation.

    b. Jamal was _____.

3. I work in a bank. The work **interests** me.

    a. It is _____ work.

    b. I am _____ in the work.

    c. The work is _____.

4. The team won the game 60 to 0. The score still **amazes** the players.

    a. The score was _____.

    b. The players are _____.

    c. It was an _____ score.

QUIZ 15 Participial Adjectives: -ed vs. -ing (Chart 10-8)

Directions: Circle the correct adjectives.

Example: Babies are (**interested**)/ **interesting** in black and white objects.

1. It's very stormy outside. The children are **frightened / frightening** by the wind and the thunder. The noise of the storm is **frightened / frightening** to them.

2. Mr. Peters gave a powerful speech. The audience was **excited / exciting** about his ideas for the future.

3. Going to the dentist is **scary / scared** for a lot of people, but I am not usually **scary / scared**. However, I never feel completely **relaxed / relaxing** at the dentist.

(continued on next page)

4. Jason just saw a new action movie, but it was really **disappointed / disappointing**. He was **surprised / surprising** that the movie got good reviews.

5. Watching a group of gorillas is **fascinating / fascinated**. They seem to have a lot of similar actions to humans. I am very **interested / interesting** in learning more about them.

QUIZ 16 Get + Adjective; Get + Past Participle (Chart 10-9)

Directions: Complete the sentences with the correct forms of *get* and the words in the box.

| cold | dark | hungry | lost | rich | sunburn |
| confuse | dirty | ~~invite~~ | nervous | serious | |

Example: The Schallers are having a party. Did you ___get invited___?

1. Every time the teacher explains a new problem, I _____. He needs to be clearer.

2. It took Sandy two hours to find our house. She said she _____ and had to ask for directions.

3. My son was playing in the sandbox all afternoon. He _____ and had to take a bath before bed.

4. Before Clement gives a speech, he always _____. His hands shake, and his mouth gets dry.

5. The sunset is beautiful tonight. It's _____ now, and soon it'll be bedtime.

6. Marcel wants to earn a lot of money and _____. He'll have to work hard.

7. Please put on some sunscreen. The last time you were in the sun, you _____.

8. You just had breakfast, and now you want a snack? You sure _____ quickly!

9. This situation isn't funny. We need to _____ and think of a solution.

10. Joe's hands always _____ when he's outside during the winter. He needs to wear gloves.

QUIZ 17 Be Used To / Be Accustomed To (Chart 10-10)

A. Directions: Complete each sentence with the correct affirmative or negative form of *be used to*.

Example: I live in Oslo. I _____*am used to*_____ snowy winters.

1. Antonio's mom uses lots of jalapeño peppers. He _____ eating spicy food.

2. Lucia lived in Florida for years. At that time, she _____ cold weather.

3. I am usually on time for meetings. I _____ being late.

4. When the Ingrahams lived in their old neighborhood, they had many Italian-speaking neighbors. They _____ hearing Italian.

5. We usually get up early. We _____ sleeping until noon.

B. Directions: Complete each sentence with the correct affirmative or negative form of *be accustomed to*.

Example: I live in Taipei. I _____*am accustomed to*_____ hot summers.

1. The basketball team practices every day. The players _____ spending a lot of time in the gym.

2. I grew up in a small town. I _____ living in a quiet place. After I moved to a big, noisy city, things changed.

3. In Japan, people usually don't tip in restaurants, so Sachiko _____ tipping.

4. Rick is taking three chemistry courses. He _____ writing lab reports.

5. I got a new smartphone last month. It was quite difficult to use at first, but now I _____ using it.

QUIZ 18 Used To vs. Be Used To (Chart 10-11)

Directions: Complete the sentences with *used to* or the correct form of *be used to* and the verb in parentheses.

Example: I (live) _____*used to live*_____ in Berlin, but now I live in Munich.

1. I love learning about other places and other cultures. When I was younger, I (travel) _____ a lot. When I traveled, I (spend) _____ time in small cities rather than busy tourist areas.

(continued on next page)

2. My husband and I like to try a variety of international recipes, so we
   (*eat*) _____ interesting dishes.

3. A: Where did you live before you moved here?

   B: I (*live*) _____ in Alaska.

   A: Alaska? Really? I'll bet that's really different from here.
   (*you, live*) _____ here now?

   B: Not really. I miss the mountains.

4. Bryan has to travel a lot for his job. He (*be*) _____ away from home several times a year. His family (*be*) _____ it, too.

5. A: What did your husband do before he became president of the company?

   B: He (*work*) _____ as an account executive.

6. This town (*have*) _____ lots of trees and parks, but now it is mostly new houses and apartments. I liked it a lot better when it had more trees. I (*be, not*) _____ all this growth.

QUIZ 19 Be Supposed To   (Chart 10-12)

A. Directions: Make sentences with similar meanings. Use the correct form of **be supposed to**.

Example: The hospital expects doctors to work weekends once a month.

   Doctors are supposed to work weekends once a month.

1. The school director expects students to wear uniforms.

   _____

2. The phone company expects customers to pay their bills on time.

   _____

3. Our mom expected us to be on time, but we weren't. We were late.

   _____

4. My doctor expected me to make an appointment, but I forgot.

   _____

5. The weather reporter expected it to snow last night.

   _____

(continued on next page)

B. Directions: Check (✓) the incorrect sentences and correct them. The first one is done for you.

1. ✓ It $\overset{is}{\wedge}$ supposed to snow tonight.

2. ____ Drivers are suppose to drive more slowly in rainy weather.

3. ____ The new grocery store is supposed to open next week.

4. ____ You not supposed to wear shoes in the house.

5. ____ We are supposed help Graciela with the dishes.

6. ____ You didn't supposed to go to the mall after school

QUIZ 20 Chapter Review

Directions: Circle the correct completions.

Example: A: Where is my piece of cake?

  B: Sorry. It ____ by the dog.

   a. eating   c. ate

   (b.) got eaten   d. is eating

1. A: Are you enjoying your time in Paris?

   B: Yes, but I'm tired. I ____ the time change yet.

   a. wasn't used to   c. am not used to

   b. didn't use to   d. am not get used to

2. A: How was the lecture at the museum?

   B: It was terrible! I was really ____.

   a. boring   c. boredom

   b. born   d. bored

3. A: Where did you get that hat?

   B: It ____ in Nepal, but I bought it online.

   a. was made   c. makes

   b. made   d. is being made

4. A: Your face is really red.

   B: I stayed in the sun so long yesterday that I ____.

   a. was sunburn   c. was getting sunburn

   b. got sunburned   d. sunburned

(continued on next page)

5. A: Why is Martina so happy?

   B: She just _____ Nicolo.

   a. got engaged to
   b. got engaged with
   c. gets engaged with
   d. was gotten engaged to

6. A: Where's your motorcycle?

   B: At the mechanic's shop. It _____.

   a. is repairing
   b. is being repaired
   c. being repaired
   d. repairing

7. A: Did you go sailing on your vacation?

   B: Yes. We had an _____ trip in the Sea of Cortez.

   a. excited
   b. exciting
   c. excite
   d. excites

8. A: You look like you're in hurry.

   B: I am. This project _____ done by 5:00 p.m.

   a. has to been
   b. will to be
   c. has to be
   d. must to being

9. A: Why are the police here?

   B: A car accident _____ a few minutes ago.

   a. is happened
   b. was happened
   c. happens
   d. happened

10. A: Do you want to go on a picnic?

    B: I don't think so. It _____ rain this afternoon.

    a. is supposed
    b. is supposed to
    c. be supposed to
    d. is suppose to

CHAPTER 10 – TEST 1

Part A *Directions:* Complete the sentences with the correct active or passive forms of the verbs in parentheses.

I. SITUATION: A Speech Contest

Yesterday, we (*have*) _____1_____ a speech contest at my high school. Students (*speak*) _____2_____ about many different topics. Their speeches (*judge*) _____3_____ by teachers. At the end of the day, the top three student speakers (*give*) _____4_____ prizes for their excellent speeches.

II. SITUATION: A Town Meeting

At the town meeting last night, there (*be*) _____5_____ a very interesting discussion. People (*ask*) _____6_____ by community leaders to discuss several issues, but the community only (*want*) _____7_____ to talk about one issue: a proposal to build a new shopping center. Business people in the audience (*argue*) _____8_____ that a shopping center would bring jobs to the town. Most other people (*say*) _____9_____ it would destroy the small-town feeling of the community. It was clear that more time (*need*) _____10_____ for discussion of the issue.

Part B *Directions:* Circle the correct answers.

1. A: What happened to the roof of your car?

    B: It _____ in the windstorm.

    a. damaged   c. been damaged

    b. got damaged   d has damaged

2. A: The baby is so tired. She can hardly keep her eyes open.

    B: She _____ to bed as soon as possible.

    a. should put   c. was put

    b. got put   d. should be put

3. A: Do you prefer coffee or tea?

    B: I _____ a lot of coffee, but now I prefer tea.

    a. used to drinking   c. am not use to drink

    b. drinking   d. used to drink

(continued on next page)

4. A: My patient _____ in Room 303.

   B: Doctor, he was moved to the second floor.

   a. is supposed to be
   c. is suppose to be

   b. supposed to
   d. was suppose to be

5. A: This painting is beautiful.

   B: Thank you. It _____ by one of our student artists.

   a. was painting
   c. was painted

   b. is painted
   d. has painted

6. A: How do you like college?

   B: I like living in the dorm, but I _____ the food.

   a. am not use
   c. didn't used to

   b. am not used to
   d. didn't use to

7. A: Why are you moving out of the city?

   B: It's getting too big for me. Too many new buildings _____.

   a. are being built
   c. is being built

   b. are building
   d. is

8. A: When does the wedding dinner begin?

   B: Dinner _____ at 7:00 P.M.

   a. served
   c. will be serving

   b. will be served
   d. is serving

Part C *Directions:* Circle the correct adjectives.

1. A: That roller coaster ride was **thrilled / thrilling!** Let's go again.

   B: You can go, but not me. It was too fast for me. I was really **frightened / frightening**.

2. A: I am **disappointed / disappointing** that the neighborhood library will close soon.

   B: Yes, that was **surprised / surprising** news. I don't understand it. It's **confused / confusing** because the library is so popular.

3. A: What's your favorite clothing store?

   B: Jasmine's of course! They have an **amazed / amazing** variety of new styles. I'm always **excited / exciting** to see what's new.

Part D *Directions:* Correct the errors.

1. Ben and Rachel were get engaged last month.
2. Many people are opposed against higher taxes.
3. I heard my name. Who was called me?
4. Dogs in the park supposed be with their owners.
5. A package was come a few minutes ago.
6. Our apartment must clean before the party next week.
7. I used to running, but now I walk for exercise.
8. The fish isn't ready yet. It should be to cook a little longer.
9. We enjoyed our time in Malaysia, but we exhaust from the heat.
10. Jorge can skateboard for hours before he gets tiring.

CHAPTER 10 – TEST 2

Part A *Directions:* Complete the sentences with the correct active or passive forms of the verbs in parentheses. More than one answer is possible.

When I (*return*) _____ from my business trip last night, I was
                        1
pleasantly surprised. My husband (*decide*) _____ to present
                                            2
me with a "welcome home" gift. The entire house (*clean*) _____.
                                                          3
The windows (*wash*) _____, and the furniture and floors
                      4
(*dust*) _____. The house had never looked so beautiful.
         5
I (*thank*) _____ him for doing such a wonderful job. For
            6
a minute, he (*look*) _____ embarrassed, and then he finally
                       7
(*say*) _____, "Honey, everything (*do*) _____ by
         8   9
a cleaning company. But I (*get*) _____ a great discount!"
                                  10

Part B *Directions:* Circle the correct answers.

1. A: Did you make your sweater?

   B: No, but it _____ by hand, and a friend gave it to me.

   a. made   c. is made

   b. was making   d. was made

2. A: Did Roberto quit his job?

   B: I _____ that he took a long vacation.

   a. telling   c. have told

   b. was told   d. tell

3. A: Who are you waiting for?

   B: My mom _____ pick me up in a few minutes.

   a. is supposed to   c. supposed to

   b. will suppose to   d. supposes to

4. A: When can I pick up the car?

   B: It _____ by tomorrow afternoon.

   a. should be fixed   c. should fixed

   b. should fix   d. should be fixing

(continued on next page)

5. A: The weather certainly is cold here.

   B: I don't think I'll ever _____ it.

   a. am used to
   b. be use to
   c. get used to
   d. got used to

6. A: Why is there such a long line of people?

   B: All passengers' passports _____ before they board the airplane.

   a. must check
   b. must to check
   c. must be checking
   d. must be checked

7. A: I love this photo of you and Dan.

   B: Thanks. It _____ when we were on vacation last summer.

   a. took
   b. is taking
   c. was taken
   d. has taken

8. A: Do you live in Wallingford?

   B: No, I _____ live there, but I moved to Shorewood last year.

   a. am used to
   b. used to
   c. use to
   d. got use to

Part C *Directions:* Circle the correct adjectives.

1. A: How was the movie?

   B: I was **disappointed / disappointing**. It was supposed to have new special effects, but they weren't new. They were just **bored / boring**.

   A: That's too bad. Would you like to see another movie tonight? I hear *Chaos* is pretty good.

   B: Sure! I'd like to see an **excited / exciting** movie.

2. A: I read a **shocked / shocking** article yesterday. Scientists say that global warming is quickly getting worse.

   B: I feel **depressed / depressing** just thinking about.

   A: We all need to change our habits to reduce global warming.

3. A: I just finished a **fascinated / fascinating** book about travel to Mars. Do you want to borrow it?

   B: No, thanks. I'm not really **interested / interesting** in space travel.

Part D *Directions:* Correct the errors.

1. The Jeffersons have been married with each other for 50 years.
2. My new boss is very interested from my work experience.
3. My husband and I used to living on a houseboat. Now we rent an apartment downtown.
4. Dr. Barry was arrived two hours late and missed the meeting.
5. That wallet makes of leather. I really like it.
6. The dog began to cross the highway, but there were so many cars that he got scare.
7. Pietro is from southern Italy. He isn't used to drive in snow.
8. The children are very excited for going to the aquarium.
9. Rita's wedding is today. She getting very nervous.
10. What time is the play supposed to be start?

CHAPTER 11   Count / Noncount Nouns and Articles

QUIZ 1 A vs. An   (Chart 11-1)

Directions: Circle *a* or *an*. The first one is done for you.

SITUATION: *Things You See in a Park*

1. a   (an)   excited child
2. a   an   jogger
3. a   an   interesting fountain
4. a   an   playground
5. a   an   butterfly
6. a   an   ice cream cone
7. a   an   bike path
8. a   an   historic monument
9. a   an   unusual insect
10. a   an   lovely flower garden
11. a   an   unique bird

QUIZ 2 A / An vs. Some   (Charts 11-1 → 11-3)

Directions: Complete the phrases with *a*, *an*, or *some*. The first one is done for you.

SITUATION: *In the Attic*

Lindsay and Lee cleaned out their attic. They found a lot of stuff! They found …

1. ____some____ dishes
2. _____ old lamp
3. _____ tennis rackets
4. _____ picture frame
5. _____ luggage
6. _____ old clothes
7. _____ ugly chair
8. _____ dusty carpet

(continued on next page)

9. _____ bicycle
10. _____ out-of-date calendar
11. _____ exercise equipment

QUIZ 3 Count and Noncount Nouns (Charts 11-2 → 11-4)

Directions: Add final *-s / -es* if possible. If not possible, write **Ø**.

Example: We need more toothpaste __Ø__. This tube is nearly empty.

1. Could you mail the letter _____ on the table? They need to go out today.
2. I hope you make a lot of progress _____ on your project.
3. I don't have enough time _____ to eat breakfast. I usually just have coffee _____.
4. Everyone wished the couple happiness _____ on their wedding day.
5. My parents always tried to offer help _____ when I had problems.
6. Michael has a lot of homework _____ tonight. He has assignment _____ in math, social studies, and Japanese.
7. Every anniversary, Giorgio gives his wife an expensive gift. She likes jewelry _____, and she especially loves ring _____.
8. The lawyer had some new information _____ about the case. He wanted to present more fact _____ to the judge.
9. Knowledge _____ comes with experience.
10. I love the sunshine _____, but this hot weather makes me really thirsty. I've had three large glass _____ of water today.

QUIZ 4 Much / Many (Charts 11-2 → 11-5)

Directions: Circle *much* or *many* and make the nouns plural as necessary. The first one is done for you.

SITUATION: *On the Internet*

1. How   much / **(many)**   website __s__ does Shelley visit each week?
2. How   much / many   time _____ does she spend online each day?
3. How   much / many   information _____ did Shelley search for when she did her research project?
4. How   much / many   friend _____ does she have online?
5. How   much / many   message _____ does she get each day?

(continued on next page)

6. How **much / many** photo _____ has she posted recently?

7. How **much / many** money _____ does she spend on music each month?

8. How **much / many** online game _____ does she play?

9. How **much / many** book _____ has she read online?

10. How **much / many** knowledge _____ has she gained?

11. How **much / many** experience _____ does Shelley have with online shopping?

QUIZ 5 A Few / A Little   (Charts 11-2 → 11-5)

Directions: Complete the sentences with *a few* or *a little* and the given nouns. Make the nouns plural as necessary.

Example: vegetable   We planted _____*a few vegetables*_____ in our garden.

1. traffic   There's _____ on the road, but it's not bad.

2. apple   Could you pick up _____ at the store? I need about three or four.

3. meat   There's _____ left over from dinner.

4. milk   There's only _____ left in the refrigerator. I need to buy more.

5. coin   I found _____ in your pocket when I did the laundry.

6. pepper   This soup is almost ready. It just needs _____.

7. vocabulary   This article has only _____ that I don't understand. I don't have to use my dictionary.

8. suggestion   The teacher gave her students _____ for writing topics.

9. cookie   If I am hungry between meals, I have _____.

10. money   I gave the kids _____ to buy snacks.

QUIZ 6 — A Lot Of, Some, Several, Many / Much, and A Few / A Little (Charts 11-2 → 11-5)

Directions: Circle the correct words or phrases for the nouns in **bold**. More than one word or phrase may be correct. The first one is done for you.

Example: (a few)   much   (several)   **people**

1. several   many   a little   **rain**
2. much   many   a little   **fun**
3. some   much   a lot of   **tests**
4. several   some   many   **information**
5. a few   a little a lot of   **help**
6. much   a lot of a little   **knowledge**
7. many   several   some   **postcards**
8. several   a few   a little   **insects**
9. some   much   a little   **stuff**
10. much   a few   some   **vocabulary**

QUIZ 7 — Nouns That Can Be Count or Noncount (Chart 11-6)

Directions: Circle the correct completions. The first one is done for you.

Example: I had many wonderful **experience /(experiences)** living in Europe.

1. Who would like **coffee / coffees?** I'll make some.
2. We've had **chicken / chickens** every night for dinner this week. Let's have something different tonight.
3. The hotel provides **iron / irons** for guests who need to press their clothes.
4. Professor Chang spends a lot of **time / times** with his students outside of class.
5. There's a lot of **light / lights** in this house because there are so many windows.
6. Emma colors her **hair / hairs** often. She always looks different!
7. Dennis should try wearing contact lenses for a while. He's always losing his **glass / glasses**.
8. We need to get **a paper / some paper** to wrap these birthday gifts.
9. The museum exhibit has several **work / works** of art by Claude Monet.
10. Professor Reed assigns a lot of **paper / papers** for his students to write.

QUIZ 8 Units of Measure with Noncount Nouns (Chart 11-7)

Directions: Complete each phrase with a unit of measure in the box. More than one answer may be correct. The first one is done for you.

| bag | bottle | box | can | jar |

| bowl | cup | glass | piece | slice |

At the store, I bought …

1. a ___can___ of tuna fish
2. a _____ of crackers
3. a _____ of juice
4. a _____ of olives
5. a _____ of soup
6. a _____ of rice

For a snack, I had …

7. a _____ of ice cream
8. a _____ of toast
9. a _____ of juice
10. a _____ of cheese
11. a _____ of tea

QUIZ 9 Understanding Articles with Count and Noncount Nouns (Chart 11-8)

Directions: Do the words in *italics* have a specific or non-specific meaning? Circle your answer. The first one is done for you.

1. *Insects* have six legs. — specific / (non-specific)
2. A few months ago, Joshua got a new *car*. — specific / non-specific
3. The *cell phone* on the table is mine. — specific / non-specific
4. I have all the *information* for the project. — specific / non-specific
5. It takes *courage* to be a firefighter. — specific / non-specific
6. A *lion* is a dangerous animal. — specific / non-specific
7. Trish prefers sunny *weather*. — specific / non-specific
8. I saw the new Sam Spion *movie* last night. — specific / non-specific
9. The *lemons* in the basket are from my lemon tree. — specific / non-specific
10. An *elephant* cannot jump. — specific / non-specific
11. Patrick and Andrea love *dogs*. — specific / non-specific

QUIZ 10 Articles: Specific vs. Non-specific (Chart 11-8)

Directions: Complete the sentences with the given nouns. Use *the* for specific statements. Don't use *the* for non-specific statements.

Example: flowers
a. _____Flowers_____ are a romantic gift.
b. _____The flowers_____ in my garden are colorful.

1. sunglasses
   a. _____ for sale at that store are too expensive.
   b. I have to wear _____ to protect my eyes.

2. bread
   a. My family loves fresh _____ for breakfast.
   b. _____ from our local bakery is delicious!

3. furniture
   a. I don't like their living room. _____ in it is old and dirty.
   b. They need to buy _____ for their living room.

4. children
   a. It's good for _____ to get a lot of exercise.
   b. _____ in that family are very active.

5. vocabulary
   a. _____ is important for learning any language.
   b. I don't understand this story. _____ is too difficult.

QUIZ 11 The vs. A / An (Chart 11-8)

Directions: Complete the conversations with *the*, *a*, or *an*.

Example: A: I'm very cold.
B: I have __a__ sweater you can wear.

1. A: What do you want for lunch?
   B: I'd like _____ bowl of tomato soup, please.

2. A: What's wrong? You look upset.
   B: _____ books for my biology class are really expensive.

3. A: I can't find my glasses.
   B: They're on _____ bookcase in the living room.

4. A: Do you have _____ accountant to do your taxes?
   B: Yes. Do you want her contact information?

(continued on next page)

5. A: It's a little noisy in here.

   B: I know. The people in _____ apartment above us are remodeling their kitchen.

6. A: Jackson is planning to take _____ course in web design next term.

   B: How exciting!

7. A: Have you ever been to New England?

   B: Yes, I have. _____ scenery in the fall is fantastic when the leaves change color!

8. A: _____ screen on my phone just went black. Now it won't turn on.

   B: Maybe it's time to get _____ new phone.

9. A: Maggie, did you have fun at _____ birthday party yesterday?

   B: Yes, Mommy. We had cake and ice cream!

QUIZ 12 More About Articles: Using *The* (Chart 11-9)

Directions: Complete the sentences with ***the*** or ***Ø***.

Example: __The__ library is on Davis Street.

1. This is only _____ second time Mark has played the piano for an audience.

2. Henry is in _____ bed with a bad cold. Poor guy!

3. Lauren and I have _____ same birthday.

4. Mom isn't home yet. She's still at _____ work.

5. Do you know _____ best way to get to downtown?

6. Please put _____ milk back in _____ refrigerator.

7. I have to talk to _____ my doctor about my headaches.

8. _____ sun is too hot. I need a hat.

9. Sara is having problems with _____ this assignment.

QUIZ 13 *The* for Second Mention (Chart 11-9)

Directions: Complete the sentences in the paragraph with *the* or *a*. The first one is done for you.

Yesterday, I decided to buy __a__(1) smartwatch as _____(2) birthday present for my husband. I went to _____(3) store at the mall, and _____(4) really kind salesperson helped me choose one. _____(5) watch was expensive, and I hoped my husband would be pleased with _____(6) present I had chosen. He was thrilled until he tried to sync _____(7) watch with his phone, and it didn't work. We went back to _____(8) store to ask about it. Both of us felt embarrassed, however, when _____(9) salesperson said that we had not set up the Bluetooth®. He showed how _____(10) Bluetooth worked and told us to come back anytime. We thanked him and left _____(11) mall. My husband is going to enjoy his gift!

QUIZ 14 Using *The* or Ø (Charts 11-8 and 11-9)

Directions: Complete the sentences with *the* or *Ø*.

Example: __The__ books you got from the library last week should be returned soon.

1. Sue prefers _____ chicken to fish.

2. It looks like _____ fruit in the bowl has gone bad.

3. _____ bats are active at night and sleep during the day.

4. Andrew needs to change jobs. _____ work he is doing now isn't very challenging.

5. _____ car parked behind my car is blocking my way. I need to find _____ owner.

6. _____ nurses need to study biology and chemistry.

7. These days, people send more email than _____ letters.

8. I went to that new Italian restaurant last night. _____ food was very good.

9. _____ chairs in our classroom are uncomfortable.

QUIZ 15 Review of A, An, Ø, and The (Charts 11-1 → 11-9)

Directions: Complete the conversations with *a, an, the,* or *Ø*.

Example: A: It's raining really hard!

B: No problem. I have __an__ umbrella you can borrow.

1. SITUATION: *In a Restaurant*

   A: May I take your order?

   B: Yes. I'd like _____ egg salad sandwich, please. Does it come with _____ chips?

   A: Yes, it does. What would you like to drink?

   B: I'll just have _____ glass of water.

2. SITUATION: *New in Town*

   A: Excuse me. I'm looking for _____ police station.

   B: It's right over there, across from _____ bank.

   A: Thanks. I appreciate _____ your help.

3. SITUATION: *Sleep Problems*

   A: My back is sore. I think we need _____ new bed.

   B: I totally agree. _____ bed we have is too soft. I prefer sleeping on _____ firm mattress.

4. SITUATION: *A Long Sleep*

   A: _____ animals are really interesting. Did you know that _____ black bear usually sleeps during the winter months?

   B: Yes, and then it wakes up in the spring. That is the longest nap in _____ world!

5. SITUATION: *Sports*

   A: I want to learn to play tennis, so I need to buy _____ tennis racket and _____ tennis balls. That's a lot of stuff!

   B: I know. _____ sports equipment can be expensive.

QUIZ 16   *The* or Ø with People and Places   (Chart 11-10)

Directions: Circle the correct completions.

Example: We plan to vacation in **the /** Ⓞ New Zealand, but we'll visit relatives in ⓣⓗⓔ **/ Ø** Philippines first.

1. Earthquakes are common on islands in **the / Ø** Pacific Ocean.
2. Joel wants to climb **the / Ø** Mount Kilimanjaro. He tried it last year, but he didn't reach the top.
3. Ivan and Natalya are on a two-week trip to **the / Ø** Europe.
4. Husam would like to visit **the / Ø** Abu Dhabi in **the / Ø** United Arab Emirates after he finishes his work in **the / Ø** Egypt.
5. **The / Ø** Caspian Sea is next to **the / Ø** Russia.
6. I grew up in a small town in **the / Ø** Rocky Mountains.
7. **The / Ø** Professor Cloke is replacing **the / Ø** Dr. Roverso for the rest of the term.
8. **The / Ø** North America consists of **the / Ø** Mexico, **the / Ø** United States, and **the / Ø** Canada.

QUIZ 17   Capitalization   (Chart 11-11)

Directions: Capitalize the words in the sentences as necessary.

Example: ~~n~~ew ~~y~~ears is the most important holiday in many ~~a~~sian countries.
           N   Y   A

1. theresa can't decide whether to study japanese or chinese.
2. where are you going for the summer break?
3. the alps are in switzerland, austria, and france.
4. we're reading shakespeare's *romeo and juliet* for our literature class.
5. the directions say to turn on fifth street, but this is park avenue.
6. last monday was my first day as a student at stanford university.
7. the mississippi river flows into the gulf of mexico.
8. i was supposed to be born in april, but i was born a month early, so my birthday is in march.
9. which instructor do you prefer: dr. costa or professor peterson?
10. math 441 is a very high-level math class.

QUIZ 18 Chapter Review

Directions: Circle the correct completions. The first one is done for you.

1. _____ communicate with each other by using sounds and body language.
    a. Dolphin
    b. A dolphins
    (c.) Dolphins
    d. The dolphins

2. How _____ do you want, a half or a full glass?
    a. many milk
    b. much milk
    c. much milks
    d. many milks

3. I found _____ about the history of my hometown on this website.
    a. a little information
    b. a few information
    c. a few informations
    d. a little informations

4. Visiting the rainforest in Brazil is _____ experience.
    a. a unique
    b. unique
    c. an unique
    d. the unique

5. I have _____ to pack before I move to my new office.
    a. lots of stuffs
    b. a lot of stuffs
    c. lot of stuff
    d. a lot of stuff

6. _____ people attended the movie preview. There were no empty seats.
    a. Many
    b. A few
    c. A little
    d. Much

7. Mrs. Kim drinks _____ of hot tea with breakfast every morning.
    a. a jar
    b. a cup
    c. a can
    d. a bottle

8. I need to study _____ new vocabulary this weekend.
    a. several
    b. a
    c. some
    d. many

9. Sammy ate two _____ of chocolate ice cream.
    a. bowl
    b. bowls
    c. piece
    d. pieces

10. My parents usually give me good _____.
    a. advices
    b. informations
    c. suggestion
    d. advice

11. Busy parents sometimes don't have enough _____ to spend with their children.
    a. time
    b. the time
    c. times
    d. he times

CHAPTER 11 – TEST 1

Part A *Directions:* Circle the correct completions.

1. I broke _____ and had to replace them.
   a. several dishes
   b. a little dishes
   c. one dishes
   d. much dish

2. We only have _____ left. I need to buy some more.
   a. a few flours
   b. much flour
   c. a little flour
   d. many flours

3. How _____ do you have for the weekend?
   a. many homeworks
   b. much homeworks
   c. much homework
   d. many homework

4. What _____! It's cooked just right.
   a. delicious fish
   b. delicious fishes
   c. a delicious fishes
   d. an delicious fish

5. Can I offer you a _____ of chocolate cake?
   a. glass
   b. slice
   c. pieces
   d. cup

6. Mario's mother gave him _____ to buy ice cream.
   a. a little dollars
   b. a few dollars
   c. few dollars
   d. little dollars

7. I need two _____ of pickles for the picnic.
   a. jars
   b. glasses
   c. boxes
   d. cans

8. Caitlin got _____ at the library.
   a. a lot of book
   b. lot of books
   c. a lots of books
   d. a lot of books

9. _____ is not something you can buy in a store.
   a. Beauties
   b. Beauty
   c. A beauty
   d. The beauty

10. _____ of other cultures often causes conflict in the world.
    a. Ignorances
    b. Some ignorances
    c. Ignorance
    d. Several ignorance

Count / Noncount Nouns and Articles

Part B *Directions:* Capitalize the words as necessary.

1. the lake is too cold for swimming. how about going to the indoor pool at mountain view park?
2. our biology class will be taught by dr. jones. he's a professor, not a medical doctor.
3. maria's parents are from mexico. she speaks spanish fluently.
4. the college directors plan to tear down brown hall and build a new library.
5. would you be interested in going on a boat trip down the colorado river? we would see part of the grand canyon.

Part C *Directions:* Complete the sentences with *a, an, the,* or *Ø.*

1. Is there _____ bank near here?
2. _____ Dr. Powell called. She wants to discuss _____ results from your medical tests.
3. I saw _____ police car. Was there _____ accident nearby?
4. _____ woman in the red hat has _____ question.
5. The Bakers have _____ daughter and _____ son. _____ son is away at college, but _____ daughter still lives at home.
6. Quick! Open _____ door. I don't want to drop this heavy box.
7. As I get older, I'm less excited about _____ birthdays.
8. Every day there are hundreds of _____ earthquakes around _____ world.
9. _____ Mount Kilauea is _____ active volcano in _____ Hawaii.
10. Look at _____ fog! I can hardly see _____ road.

Part D *Directions:* Correct the errors.

1. Let's get a drink of water. I'm a thirsty.
2. A weather at the beach is beautiful in August.
3. Your hair looks great. Did you get haircut?
4. For breakfast, Thomas ordered two eggs and two toasts.
5. Here's a map of United States. Do you see California?
6. I need a few more time to finish my test.
7. There are no fishes in the Dead Sea.
8. Aunt Betsy and Uncle Wes are moving to the London next month.
9. I need to have my the car checked soon.
10. Much student at Shorewood High School study Japanese.

CHAPTER 11 – TEST 2

Part A *Directions:* Circle the correct completions.

1. _____ with computer programming is necessary for this job.
    a. Experiences
    b. Experience
    c. An experience
    d. Some experiences

2. The weather forecast said there wouldn't be _____ thunder, but it was quite loud last night.
    a. a few
    b. several
    c. many
    d. much

3. Here's a _____ of chicken soup. It should help you feel better.
    a. bottle
    b. bag
    c. bowl
    d. box

4. Dr. Rodriguez tried to give her patient _____, but she wouldn't listen.
    a. some advices
    b. an advice
    c. a little advice
    d. many advices

5. Professor Johnson encourages his students to ask _____.
    a. questions
    b. lot of questions
    c. some question
    d. several question

6. There's only _____ in our garden, so tomatoes don't grow well.
    a. a little sunlight
    b. a few sunlight
    c. a few sunlights
    d. a little sunlights

7. Many doctors believe _____ can help a sick person heal more quickly.
    a. the laughter
    b. many laughters
    c. several laughters
    d. laughter

8. The sale at the toy store was very popular. At the end of the sale, there weren't _____ toys left.
    a. many
    b. much
    c. one
    d. a little

9. I'm thirsty. _____ would be nice.
    a. Glass of water
    b. A glass of water
    c. Glasses of waters
    d. Some glass of water

10. _____ can come from simple pleasures in life, such as watching a sunset.
    a. Many happinesses
    b. Much happinessed
    c. Much happiness
    d. Many happiness

Count / Noncount Nouns and Articles

Part B *Directions:* Capitalize the words as necessary.

1. the assignment for our literature class is to read the first chapter of shakespeare's *hamlet*.
2. i heard that my neighbors plan to visit england in may.
3. tomorrow there will be a concert at washington park near broadway avenue. a music group from south africa will be playing.
4. there is a miami university in ohio, but the city of miami is in florida. isn't that strange?
5. when did william begin working for the songlin corporation?

Part C *Directions:* Complete the sentences with *a, an, the,* or *Ø*.

1. Monday is _____ holiday for _____ students and _____ government employees.
2. Oops. It looks like our waiter made _____ error on our bill.
3. What's _____ difference between _____ hotel and _____ bed and breakfast?
4. _____ cost of gasoline could rise this summer.
5. _____ fish need _____ oxygen to breathe.
6. It can be dangerous to climb _____ mountains in _____ warm weather.
7. What brings more happiness: _____ health or _____ wealth?
8. We're lost. We should look at _____ map. I'm sure _____ map will have the street we're looking for.
9. We painted _____ walls of our apartment. Now we need to paint _____ ceiling.
10. _____ windows in the house were large, so the rooms were filled with _____ sunlight.

Part D *Directions:* Correct the errors.

1. I don't need a help now, but I will later on.
2. There are a little people at work who don't like the new manager.
3. The water is necessary for life.
4. I married my a brother's best friend from college.
5. Antoine reached the top of the Mount McKinley in Alaska yesterday.
6. My friends bicycled through Sahara Desert last summer.
7. Nick worked on his car for a hour, but he couldn't fix it.
8. Barcelona is in the Spain, but it's not the capital city.
9. Carlos texted his dad several time, but he didn't get a response.
10. Adrianna has so much a homework that she stays up late every night.

CHAPTER 12 Adjective Clauses

QUIZ 1 Using *Who* and *That* to Describe People   (Charts 12-1 and 12-2)

A. Directions: Add *who* or *that* to each sentence as necessary. In some cases, both answers are correct.

Example: A photographer is someone ∧ takes pictures.   *who / that*

1. Many tourists visit New York City go to Central Park.
2. The French man was in my English class loved coffee and chocolate.
3. Tobias thanked the nurse took care of him in the hospital.
4. I feel happy around people are positive about life.
5. When Maja was on the bus, she sat next to a woman was talking on her cell phone.

B. Directions: Change the "b" sentences to adjective clauses. Combine each pair of sentences using *who* or *that*.

Example: a. There's the little boy.   b. He lost his balloon in the wind.
   There's the little boy who lost his balloon in the wind.

1. a. I heard about a teenaged boy.   b. He takes gifts to children in hospitals.

2. a. Tomas met a marine biologist.   b. She once swam with sharks.

3. a. The people have fire drills twice a month.   b. They work on this boat.

4. a. The police helped an old man.   b. He was confused and lost.

5. a. The scientists are very famous.   b. They discovered DNA.

Adjective Clauses   195

QUIZ 2 Who vs. Whom in Adjective Clauses (Charts 12-2 and 12-3)

Directions: Complete the sentences with ***who*** or ***whom***. Use ***whom*** for object pronouns.

SITUATION: A Party at Our Apartment

Example: The neighbor ___whom___ I asked for help was very friendly.

1. My sister _____ lives nearby helped us prepare the food.
2. One neighbor _____ I didn't know well brought her guitar.
3. One of the people _____ came to the party was my coworker, Pat.
4. Pat is a person _____ others like immediately.
5. The police officer _____ is moving into the apartment next to us came by for a few minutes.
6. A woman _____ we met last year gave us some flowers.
7. One of the people _____ we know well told funny stories all evening.
8. The man _____ brought cookies owns a bakery.
9. My friend _____ plays in the orchestra couldn't come. She was out of town.
10. The guests _____ we invited had a good time, and so did we.

QUIZ 3 Using Who, That, Ø, and Whom to Describe People (Charts 12-2 and 12-3)

Directions: Circle all the correct completions. More than one answer is possible. The first one is done for you.

1. We liked the singers ____ gave the concert.
   (a.) who   (b.) that   c. Ø   d. whom

2. The children ____ attend Lake Forest Academy come from wealthy families.
   a. who   b. that   c. Ø   d. whom

3. The driver ____ I helped was very upset about the accident.
   a. who   b. that   c. Ø   d. whom

4. We met the firefighters ____ saved our house from burning.
   a. who   b. that   c. Ø   d. whom

5. Do you trust the person ____ Mr. Wilcox hired?
   a. who   b. that   c. Ø   d. whom

6. Students ____ come to class early can get extra help from the teacher.
   a. who   b. that   c. Ø   d. whom

7. Where is the woman ____ manages this apartment building?
   a. who   b. that   c. Ø   d. whom

8. The college student ____ I drive to school every morning is my neighbor.
   a. who   b. that   c. Ø   d. whom

(continued on next page)

9. The babysitter ____ the kids like the best is friendly and funny.
   a. who   b. that   c. Ø   d. whom

10. Are you the pharmacist ____ my doctor recommended?
    a. who   b. that   c. Ø   d. whom

11. I know a man ____ has nine brothers and sisters.
    a. who   b. that   c. Ø   d. whom

QUIZ 4 Using *That* and *Which* to Describe Things (Chart 12-4)

A. Directions: Circle the correct completions. In some cases, both answers are correct. The first one is done for you.

1. I just found out that the toys ____ were on sale are dangerous for children.
   (a.) that / which   b. Ø

2. Andreas studies languages ____ are no longer spoken.
   a. that / which   b. Ø

3. My grandparents live in the house ____ they built 50 years ago.
   a. that / which   b. Ø

4. The photographs ____ my father takes are displayed at art shows.
   a. that / which   b. Ø

5. We're having lunch at a restaurant ____ has an amazing view of the city.
   a. that / which   b. Ø

6. The college ____ George attends offers scholarships to 80 percent of its students.
   a. that / which   b. Ø

B. Directions: Change the "b" sentences to adjective clauses. Combine each pair of sentences with *which* or *that*.

Example: a. Many people like technology.   b. It makes their lives easier.
   Many people like technology which / that makes their lives easier.

1. a. William got a new winter coat.   b. It keeps him really warm.
   _____

2. a. The puppy is black and white.   b. It has a lot of energy.
   _____

3. a. The train is never late.   b. It arrives at exactly 6:53 P.M.
   _____

4. a. The ancient city is popular with tourists.   b. It was discovered in 1748.
   _____

5. a. Every day, I wear gold earrings.   b. They were a gift from my husband.
   _____

QUIZ 5 Review of Adjective Clauses (Charts 12-1 → 12-4)

Directions: Change the "b" sentences to adjective clauses. Combine each pair of sentences using **who, whom,** or **which**.

Example: a. There is the man.   b. He found our dog in the park.
   There is the man who found our dog in the park.

1. a. I spoke with an amazing woman.   b. She is smart, strong and beautiful.

2. a. Here is the new book.   b. You asked me to order it.

3. a. The young man has a broken leg.   b. He crashed into a tree.

4. a. The cell phone has all the newest features.   b. I bought it yesterday.

5. a. The kind man repaired my car for free.   b. He owns the gas station.

6. a. I don't know the student.   b. She wrote an article for the newspaper.

7. a. The mathematics professor is going to retire next month.   b. I met him last year.

8. a. The documentary was fascinating.   b. We watched it last night.

9. a. The elderly woman has no family.   b. She lives in the apartment next to mine.

10. a. The oranges are really juicy.   b. They are in the bowl on the table.

QUIZ 6 Singular and Plural Verbs in Adjective Clauses (Chart 12-5)

Directions: Circle the correct completions.

Example: The students who **studies / (study)** regularly usually get higher grades.

1. The email messages that I **gets / get** from my grandfather are funny and full of news.
2. Ethan has two kids that **plays / play** basketball after school every day.
3. The main character in my favorite TV series is a woman who **writes / write** for a comedy show.
4. Many people who **works / work** the night shift have sleep problems.
5. The stories which my children **likes / like** the best have happy endings.
6. Do you know the man who **is / are** directing traffic?
7. The woman who **sells / sell** flowers on the corner is getting married to one of her customers.
8. The house which my parents **is / are** designing uses solar power.
9. I know a man who **spends / spend** every summer sailing the Pacific Ocean.
10. Most of the people who **serves / serve** on the town council work hard.

QUIZ 7 Prepositions in Adjective Clauses (Chart 12-6)

Directions: Complete the sentences with appropriate prepositions. Then underline the adjective clauses.

Example: The chair <u>that you are sitting</u> *in* looks comfortable.

1. The music that we listened _____ was recorded in front of a live audience.
2. The professor that I spoke _____ is not available to teach the class next semester.
3. The company that I work _____ treats its employees well.
4. The people who I depend _____ the most are my parents.
5. Don't tell me about your plumbing problems. The person whom you should complain _____ is the building manager.
6. The movie theater we went _____ had a huge screen.
7. The young man whom Clara was waiting _____ is a professor at Oxford University.
8. The university that Josh graduated _____ is in Seattle.
9. The elderly man whom Marta talked _____ told stories about his life.
10. The person whom Greg lives _____ has both M.D. and Ph.D. degrees.

QUIZ 8 Prepositions in Adjective Clauses (Chart 12-6)

Directions: Complete each sentence with the information in the given sentence.

Example: The taxi is coming. I am waiting for it.

    a. The taxi for __which I am waiting__ is coming.

    b. The taxi that __I am waiting for__ is coming.

1. The radio station has 24-hour news. We listen to it.

    a. The radio station to _____ has 24-hour news.

    b. The radio station that _____ has 24-hour news.

2. The manager drives a sports car. Sebastian works for him.

    a. The manager for _____ drives a sports car.

    b. The manager whom _____ drives a sports car.

3. The school specializes in dance and drama instruction. I told you about it.

    a. The school about _____ specializes in dance and drama instruction.

    b. The school that _____ specializes in dance and drama instruction.

4. The email was in my spam folder. I was looking for it.

    a. The email that _____ was in my spam folder

    b. The email for _____ was in my spam folder.

5. The new student is from India. I spoke to her after class.

    a. The new student to _____ is from India.

    b. The new student whom _____ is from India.

QUIZ 9 Using *Whose* in Adjective Clauses (Chart 12-7)

Directions: Change the "b" sentences to adjective clauses. Combine each pair of sentences with *whose*.

Example: a. I know the teacher.   b. His class is putting on a play.

      I know the teacher whose class is putting on a play.

1. a. The little girl was sad for days.   b. Her doll was stolen.

2. a. I'm friends with a woman.   b. Her daughter is training to be a professional boxer.

3. a. In Kansas City, I met a man.   b. His parents know my grandparents.

4. a. I have a friend.   b. Her sailboat is also her home.

5. a. I enjoyed meeting the couple.   b. Their children go to the same school as our children.

6. a. The people are upset.   b. Their car was just hit.

7. a. The couple has a new baby.   b. We rent their vacation home.

8. a. I know a woman.   b. Her work involves designing home security systems.

9. a. A writer spoke about his experiences.   b. His new book is about mountain climbing.

10. a. The dog is being cared for by the staff.   b. His owner left him outside a restaurant.

QUIZ 10 Chapter Review

Directions: Correct the errors.

Example: I can't stop reading the book I started ~~it~~ last night.

1. The family whose arrived late discovered they had missed the wedding.
2. A neighbor who his son works for an airline can fly anywhere quite cheaply.
3. I ran into a man he was my boss 20 years ago.
4. Those are the students they volunteer to clean up parks on weekends.
5. The cookies are burned which I baked.
6. The woman which I see every day on the bus talks the entire time.
7. I work with a doctor which has clinic hours two evenings a week.
8. Here is the magazine has the story about home theater systems.
9. Aiko and Yutaka moved into the apartment which are on the top floor.
10. The people who dog bit the delivery man had to pay the doctor bills.

CHAPTER 12 – TEST 1

Part A *Directions:* Use the "b" sentences as adjective clauses. Combine the sentences with *who, whom, whose,* or *which*.

1. a. The mail was addressed to our neighbor. b. He lives in the apartment downstairs.

   _____

2. a. I work with a man. b. His wife trains police dogs.

   _____

3. a. The garden looks healthy again. b. I had forgotten to water it.

   _____

4. a. The pianist teaches my children piano. b. She plays in the hotel lobby on weekends.

   _____

5. a. The manager treats me fairly. b. I work for her.

   _____

6. a. A real estate agent called. b. His name is Mike Hammer.

   _____

Part B *Directions:* Complete the sentences. Use *who, that, Ø, whose, which,* or *whom*. Write all possible choices.

1. The coin _____ Karl found is very valuable.

2. I met a woman _____ has two sets of twins.

3. Let's choose a movie _____ the whole family can watch.

4. Yesterday, I spoke with a man _____ brother is a psychologist for animals.

5. The reporter _____ I had the interview with works for *The Times*.

Part C *Directions:* Circle the correct completions.

1. Have you met the people who **is / are** renting that house?

2. An astronomer is a scientist who **studies / study** stars and planets.

3. There are several students in my class whose families **lives / live** in other countries.

4. The worker who **is / are** meeting with the supervisor is going to lose his job.

5. These are the shoes that **is / are** the most comfortable.

Part D *Directions:* Complete the sentences with appropriate prepositions.

1. The woman that I was telling you _____ is the president of the company.
2. The hostel that we stayed _____ wasn't expensive.
3. Rebecca is an employee whom other workers can always depend _____.
4. Cambridge is the university that Omar graduated _____.
5. The person who I am waiting _____ is often late.
6. Jeremy Graydon is a singer whom I am never tired of listening _____.

Part E *Directions:* Complete each sentence with the information in the given sentence.

1. The photographs were amazing. He paid a lot of money for them.
   a. The photographs that _____ were amazing.
   b. The photographs for _____ were amazing.
2. The woman was late. A taxi was waiting for her.
   a. The woman for _____ was late.
   b. The woman _____ for was late.
3. I spoke to a police officer. She was very tall.
   a. The police officer that _____ was very tall.
   b. The police officer to _____ was very tall.

Part F *Directions:* Correct the errors.

1. The firefighters who they put out the fire were dirty and exhausted.
2. The sofa we ordered it still hasn't arrived.
3. The man that for I work is blind.
4. Here is the receipt which you asked.
5. The finger is healing well which I broke.
6. I studied with a professor who his books are well known around the world.
7. I met a little girl whose favorite food are mushrooms.

CHAPTER 12 – TEST 2

Part A *Directions:* Use the "b" sentences as adjective clauses. Combine the sentences with *who, whom, whose,* or *which.*

1. a. The earphones didn't work. b. I bought them.

   _____

2. a. The couple was surprised. b. Their horse won the race.

   _____

3. a. The supervisor is very kind. b. People like to work for her.

   _____

4. a. We met a boy. b. His dog can do a lot of tricks.

   _____

5. a. The actor won an Academy Award. b. He starred in several movies last year.

   _____

6. a. The little girl picked some of the flowers. b. They grow in my garden.

   _____

Part B *Directions:* Complete the sentences with *who, that, Ø, whose, which,* or *whom.* Write all possible choices.

1. The sandwiches _____ you made were delicious.

2. The boy _____ we hired to do yard work is saving for a trip to Nepal.

3. The book _____ you are reading has been translated into several languages.

4. I went to school with a man _____ software company has made him very rich.

5. The dress _____ Jasmine wore to the job interview had a red belt.

Part C *Directions:* Circle the correct completions.

1. My husband and I have a friend who **designs / design** jewelry for movie stars.

2. That is the bike which Michelle **wants / want** for her birthday.

3. Where are the socks that **goes / go** with those pants?

4. I study with a professor who **speaks / speak** several languages fluently.

5. The neighbors whose dog **barks / bark** all night are rarely home.

Part D *Directions:* Complete the sentences with appropriate prepositions.

1. This is the podcast that everyone at school is listening _____.
2. The woman who Paul introduced me _____ is a teacher.
3. My neighbor waved _____ me from across the street.
4. Hans is a financial advisor whom you should complain _____.
5. These are the books that I told you _____.
6. The university that Bridget went _____ is M.I.T.

Part E *Directions:* Complete each sentence with the information from the given sentence.

1. My nephew lives in Argentina. I told you about him.

    a. My nephew that _____ lives in Argentina.

    b. My nephew about whom _____ lives in Argentina.

2. Meg found the earrings. She had been looking for them.

    a. Meg found the earrings that _____.

    b. Meg found the earrings for _____.

3. The artist is speaking at 7:30 P.M. You introduced me to her.

    a. The artist to _____ is speaking at 7:30 P.M.

    b. The artist that _____ is speaking at 7:30 P.M.

Part F *Directions:* Correct the errors.

1. The pillows Sandra had on her bed was too hard.
2. The train it came through the tunnel blew its whistle.
3. The suitcase has lots of pockets which I bought.
4. The doctor whose operated on my father is very skilled.
5. I work with a woman grew up in the same neighborhood as me.
6. The taxi driver which drove us to the hotel charged too much.
7. Here's an article that you might be interested.

CHAPTER 13 Gerunds and Infinitives

QUIZ 1 Verb + Gerund (Chart 13-1)

Directions: Complete each sentence with the correct form of a verb in the box. The first one is done for you.

~~bake~~	do	exercise	open	travel	work
buy	drive	move	run	wash	

1. I always enjoy _____baking_____ cakes.
2. Would you mind _____ the door for me? My hands are full.
3. Hiro put off _____ his homework all weekend. Now he is too tired to do it.
4. My husband and I are thinking about _____ to Asia next year.
5. Will's car was making a strange noise, but he kept on _____ because he didn't want to stop on the freeway.
6. We finished _____ the windows, and now we can see through them clearly.
7. The Davisons have discussed _____ closer to the city so Mr. Davison would have a shorter commute to work.
8. Victor quit _____ when he turned 65, and he is really enjoying retirement.
9. I always consider _____ at the gym after work, but usually I'm too lazy to work out.
10. We are going to postpone _____ a new car until we save more money.
11. Katarina and Liz are talking about _____ in the city marathon next year.

QUIZ 2 Go + -ing (Chart 13-2)

Directions: Complete the sentences with the correct form of **go** and the verbs in parentheses.

Example: I (swim) __go swimming__ at Flathead Lake every summer.

1. Tomorrow, I (shop) _____ for new summer clothes.
2. Last weekend, we (sail) _____ on our new sailboat.
3. Let's (camp) _____ this weekend. The weather is supposed to be quite warm.
4. For Mike's 75th birthday, he (skydive) _____. His wife thought he was crazy.
5. Mr. Blake knows a great river nearby where we can (fish) _____.
6. To stay in shape, Janet (jog) _____ every morning before work.
7. In Japan, it's popular to (hike) _____ on Mount Fuji.
8. My classmates and I (bowl) _____ next weekend.
9. Yesterday we (sightsee) _____ with our friends who are visiting from Spain.
10. Paolo usually (dance) _____ with his friends on Saturday nights.

QUIZ 3 Verb + Gerund or Infinitive (Charts 13-1, 13-3 and 13-4)

Directions: Complete the sentences with the gerund or infinitive form of **work**.

Example: Martin can't stand __working__ at his current job. He wants to quit.

1. Hans has talked about _____ for a computer animation company.
2. Charles pretends _____ when his boss is around.
3. Dr. Bennett enjoys _____ at the children's hospital.
4. The teenagers discussed _____ at part-time jobs during the summer.
5. Jae plans _____ in his father's company after he graduates.
6. Mari hopes _____ as a flight attendant someday.
7. The builders postponed _____ on the new bridge because of a storm.
8. Elena might consider _____ as a reporter downtown.
9. Mrs. Russo intends _____ until she is 65.
10. My brother expected _____ last night, but he didn't have to.

QUIZ 4 Verb + Gerund or Infinitive (Charts 13-1 → 13-4)

Directions: Circle the correct completions. In some cases, both answers are correct.

Example: Pat hates ____ sad movies.

    (a.) to watch   (b.) watching

1. Stan offered ____ Christopher with his math.

    a. to help   b. helping

2. In the mountains, it continued ____ for a week.

    a. to snow   b. snowing

3. We have some free time before dinner. Let's go ____.

    a. to sightsee   b. sightseeing

4. Jerry was supposed ____ his uncle at the train station, but he forgot.

    a. to meet   b. meeting

5. Have you thought about ____ extra employees for the holiday season?

    a. to hire   b. hiring

6. Miroslav and Ivana have decided ____ at the same company.

    a. to work   b. working

7. Pierre began ____ English six months ago.

    a. to study   b. studying

8. Mrs. Allen can't stand ____ out in icy weather.

    a. to go   b. going

9. We can't wait ____ our cousins on the coast.

    a. to visit   b. visiting

10. Bobby learned how ____ a computer when he was five years old.

    a. to use   b. using

11. I would like ____ on those shoes, please. I need size 6.

    a. to try   b. trying

12. Mr. and Mrs. Rowe enjoy ____ ice cream on summer evenings.

    a. to eat   b. eating

(continued on next page)

13. Would you mind _____ me some money?

    a. to lend   b. lending

14. The supervisor agreed _____ the employees extra for working on the weekend.

    a. to pay   b. paying

15. The little boy started _____ when he couldn't find his mother.

    a. to cry   b. crying

16. Emma seems _____ in a hurry this morning.

    a. to be   b. being

17. Charlie puts off _____ laundry until all his clothes are dirty.

    a. to do   b. doing

18. When my best friend moved away, she promised _____ me often.

    a. to call   b. calling

19. Brad and Susan discussed _____ across Russia by train.

    a. to travel   b. traveling

20. I wanted _____ my homework before I went to bed, but I was too tired.

    a. to finish   b. finishing

QUIZ 5 Preposition + Gerund (Chart 13-5 and Appendix C-2)

Directions: Complete the sentences with the correct prepositions.

Example: He is responsible ___*for*___ paying the bills.

1. Thank you _____ coming.
2. Jay is afraid _____ swimming in the ocean.
3. I believe _____ having money in a savings account.
4. He isn't nervous _____ getting married.
5. Please plan _____ coming for dinner.
6. We look forward _____ seeing you next month.
7. The kids are excited _____ flying to Los Angeles.
8. I feel _____ going shopping today.
9. Maya is good _____ drawing cartoons.
10. I apologize _____ hurting your feelings.

QUIZ 6 By vs. With (Chart 13-6)

Directions: Complete the sentences with *by* or *with*.

Example: The package was delivered to the wrong apartment ___by___ mistake.

1. We traveled around the country last summer _____ train.
2. At my office, people usually communicate with each other _____ email.
3. The salesperson greeted us _____ a smile.
4. The barber cut Joe's hair _____ special scissors.
5. It's easy to get around Tokyo _____ subway.
6. The nurse took the patient's temperature _____ a thermometer.
7. My sister lives far away, but we stay in touch _____ phone or email.
8. You can pay _____ credit card, but not _____ check.
9. Marcos measured the paper _____ a ruler.

QUIZ 7 By + Gerund (Chart 13-6)

Directions: Complete each sentence with *by* and the correct form of a verb in the box. The first one is done for you.

change	exercise	look	save	speak	work
do	~~have~~	make	send	wash	

1. Nina's friends surprised her ____by having____ a birthday party for her.
2. You can help me _____ the windows that I can't reach.
3. We can find out the price of the movie tickets _____ online.
4. Andy was able pay for his motorbike _____ money from his part-time job.
5. Students can improve their grades _____ extra work.
6. I finally reached Julien _____ him a text message.
7. The teacher calmed the crying child _____ in a quiet, gentle voice.
8. Nina keeps her email secure _____ her password every three months.
9. Khalid stays in shape _____ three times a week.
10. Kwon has gotten a lot of experience _____ at several different jobs.
11. The magician entertained the children _____ things disappear.

QUIZ 8 Gerunds as Subjects and *It* + Infinitive (Charts 13-7 and 13-8)

A. Directions: Make sentences with the same meaning. Use a gerund as the subject.

SITUATION: *Preparing for an Exam*

Example: It is helpful to organize all your study materials.

    Organizing all your study materials is helpful.

1. It is necessary to have a quiet place to study.

2. It is fun to study with friends.

3. It is difficult to learn a lot of new information.

4. It is a good idea to take short breaks.

5. It is important to get a good night's sleep.

B. Directions: Make sentences with the same meaning. Use *It* and an infinitive. Use *for (someone)* as needed.

SITUATION: *Taking Care of Our Planet*

Example: Predicting the future of life on this planet is impossible.

    It is impossible to predict the future of life on this planet.

1. Recycling everything you can is important.

2. Using public transportation instead of your car is helpful.

3. Turning off lights when you leave a room is sensible.

4. Saving water is possible for everyone.

5. Taking care of our planet is necessary for all of us.

QUIZ 9 Expressing Purpose: *In Order To* (Chart 13-9)

A. Directions: Add *in order* where possible.

Example: Bill called the drugstore˄to ask a question. *in order*

Alison would prefer to stay home tomorrow. *no change*

1. Kim applied to Whitman University for next year.
2. Judy is moving to her hometown to be closer to her elderly parents.
3. Tom got new glasses to see better.
4. I have to be sure to pay my bills today.
5. Frankie practiced driving almost every day to pass the driving test.

B. Directions: Answer the questions with *in order to* and the words in parentheses.

Example: Why did you go on a diet? (*lose weight*)
   I went on a diet in order to lose weight.

1. Why did you turn down the TV? (*hear you better*)

   _____

2. Why did you wear socks to bed? (*keep my feet warm*)

   _____

3. Why did you take out a loan? (*buy a new car*)

   _____

4. Why did you call the doctor? (*make an appointment*)

   _____

5. Why did you turn off your phone? (*get some sleep*)

   _____

QUIZ 10　Expressing Purpose: *To* vs. *For* (Chart 13-9)

Directions: Complete the sentences with *to* or *for*.

Example: Last year, I went to Europe ___*to*___ visit a college friend.

Last weekend, I went to the mountains …

1. _____ ski with friends.
2. _____ a ski trip.
3. _____ have fun.
4. _____ spend time away from the city.
5. _____ some fresh air and relaxation.

Yesterday, I made an appointment …

6. _____ my husband.
7. _____ see our lawyer.
8. _____ a meeting with our lawyer.
9. _____ speak with our lawyer.
10. _____ get some legal advice.

QUIZ 11 Using Infinitives with *Too* and *Enough* (Chart 13-10)

Directions: Complete the sentences with the correct form of the given adjectives and *too* or *enough*.

Example: What did you think of that movie?

    a. good   It was _____*good enough*_____ to watch again.

    b. sleepy   I was _____*too sleepy*_____ to enjoy it.

1. It's hot outside. I can't work in the garden.

    a. hot   It's _____ to work in the garden.

    b. cool   It isn't _____ to work in the garden.

2. I can't eat plain yogurt.

    a. sour   Plain yogurt is _____ for me to eat.

    b. sweet   Plain yogurt isn't _____ for me to eat.

3. I'm not going to make an omelet for breakfast.

    a. eggs   I don't have _____ to make an omelet.

    b. tired   I'm _____ to cook breakfast.

4. The brakes on the car are bad. Don't drive it.

    a. safe   The car isn't _____ to drive.

    b. dangerous   The car is _____ to drive.

5. Your shirt still has ink spots on it. You can't wear it.

    a. clean   The shirt isn't _____ to wear.

    b. dirty   It's _____ to wear.

QUIZ 12 Chapter Review

Directions: Circle the correct completions. In some cases, both answers may be correct.

Example: I don't enjoy ____ to parties.

    a. to go   (b.) going

1. We plan ____ a new computer soon.

    a. to get   b. getting

2. Jason learned how ____ when he was three years old.

    a. to ski   b. skiing

3. Although Yasuko is 15, she is afraid of ____ alone in her house.

    a. to stay   b. staying

4. Professor Dunn always wears a suit and tie to class ____ professional.

    a. to look   b. looking

5. I wanted ____ you earlier, but my phone battery died.

    a. to call   b. calling

6. Several students continued ____ during the class even after the teacher asked them to stop.

    a. to talk   b. talking

7. Tony believes in always ____ honest with people.

    a. to be   b. being

8. He's afraid ____ in elevators.

    a. to ride   b. riding

9. We'd love ____ together with you during the holidays.

    a. to get   b. getting

10. Tina doesn't mind ____ on holidays because she gets paid overtime.

    a. to work   b. working

11. Don't put off ____ your homework until late at night.

    a. to do   b. doing

(continued on next page)

12. Our math tutor promised ____ us a break after we solved seven algebra problems.

    a. to give   b. giving

13. It's important for children ____ the value of money.

    a. to learn   b. learning

14. Thank you for ____ with the chemistry project.

    a. to help   b. helping

15. Kris and Sophie arrived at the theater early ____ good seats.

    a. to get   b. getting

16. The cat was caught on a high branch of a tree. Andre rescued her by ____ the tree.

    a. climb   b. climbing

17. The students searched the internet ____ information for their research projects.

    a. to find   b. finding

18. You need patience ____ young children.

    a. to teach   b. teaching

19. Geoff likes ____ to music at night because it helps him fall asleep.

    a. to listen   b. listening

20. Grace keeps her dog in a fenced area ____ him from running away.

    a. to stop   b. stopping

CHAPTER 13 – TEST 1

Part A *Directions:* Complete the sentences with the gerund or infinitive form of the verbs in parentheses. Add prepositions where necessary.

1. A young woman called (*get*) _____ help when she had car trouble on the highway.

2. I enjoy (*read*) _____ a few good magazines when I need (*relax*) _____ .

3. My daughter promised (*call*) _____ us as soon as she gets her examination results. She hopes (*get*) _____ into medical school. She really wants (*become*) _____ a surgeon.

4. Joe insists (*get*) _____ to the airport at least five hours before his flight. He refuses (*arrive*) _____ any later than that. (*Be*) _____ in control relaxes him, and he's much calmer about flying.

5. The children would like (*build*) _____ a snowman today. It began (*snow*) _____ last night, and now the snow is quite deep. They plan (*make*) _____ the snowman as tall as the tree outside their house. School has been canceled, and they are excited (*have*) _____ some time off. When they finish (*build*) _____ their snowman, they're going to go (*skate*) _____ near their house.

Part B *Directions:* Complete the sentences with the correct prepositions.

1. Mr. Thomas is responsible _____ planning the meeting, so he's a little nervous.

2. Brad and Heidi hadn't seen each other since high school. They met at the airport _____ chance.

3. Becky is really good _____ baking bread.

4. After working hard all year, Alex and Donna are looking forward _____ having a family vacation.

5. Carlos takes a cycling trip once a year because he thinks traveling _____ bike is the best way to see the countryside.

6. The mother comforted her son _____ holding him very close for a few minutes.

7. I was going to cut the lettuce _____ a knife, but my sister showed me it was faster to do it _____ a pair of scissors.

8. I am interested _____ going to a private university if I can get a scholarship. I'm excited _____ studying business management.

Part C *Directions:* Complete the sentences with the correct form of the given adjectives and *too* or *enough*.

1. Jenny is sick. She can't go to school.

   a. sick   Jenny is _____ to go to school.

   b. healthy   Jenny isn't _____ to go to school.

2. Wait a few minutes before you touch the pan. It's still hot.

   a. cool   The pan isn't _____ to touch.

   b. hot   The pan is _____ to touch.

3. This chocolate tastes terrible. It has no sugar in it, so it's very bitter.

   a. sweet   This chocolate isn't _____ to eat.

   b. bitter   This chocolate is _____ to eat.

Part D *Directions:* Check (✓) the correct sentence in each pair.

1. a. ____ It is necessary to practice English as much as possible for students.

   b. ____ It is necessary for students to practice English as much as possible.

2. a. ____ We can continue discussion this topic tomorrow.

   b. ____ We can continue discussing this topic tomorrow.

3. a. ____ I enjoy walking because it is good exercise.

   b. ____ I enjoy to walk because it is good exercise.

4. a. ____ Playing golf it is a popular pastime for many people.

   b. ____ Playing golf is a popular pastime for many people.

5. a. ____ If we arrive at the resort before noon, we can go hiking in the mountains.

   b. ____ If we arrive at the resort before noon, we can go to hiking in the mountains.

6. a. ____ Toshi was surprised that the salesclerk apologized being rude.

   b. ____ Toshi was surprised that the salesclerk apologized for being rude.

7. a. ____ Jamie unlocked her car by the electronic key.

   b. ____ Jamie unlocked her car with the electronic key.

8. a. ____ If we hurry, we still have time to go swimming in the lake before dark.

   b. ____ If we hurry, we still have time to go to swim in the lake before dark.

9. a. ____ Helen enjoys to be the center of attention at a party.

   b. ____ Helen enjoys being the center of attention at a party.

CHAPTER 13 – TEST 2

Part A *Directions:* Complete the sentences with the gerund or infinitive form of the verbs in parentheses. Add prepositions where necessary.

1. Colin quit (*work*) _____ as a shipping clerk two weeks ago. He expects (*start*) _____ a new job as a document specialist in about a month.

2. We can't afford (*go*) _____ on an expensive vacation this year, so we're going camping.

3. Pedro is good (*kick*) _____ a soccer ball with his left foot even though he is right-handed. (*Play*) _____ soccer is his favorite free-time activity.

4. Bill doesn't mind (*listen*) _____ to classical music, but he prefers rock.

5. When she gets home from work, Liz doesn't feel like (*cook*) _____ . She just wants (*rest*) _____ on the sofa with a good book or magazine. She often puts off (*make*) _____ dinner, but her family doesn't mind (*eat*) _____ a little later.

6. I have always dreamed (*visit*) _____ Paris. When I go there, I want to go (*sightsee*) _____ and see the Eiffel Tower, the Arc de Triomphe, and other famous sites. I am a little worried (*speak*) _____ French. I tried (*learn*) _____ French online, but I it was difficult and I don't speak well. In any case, when I decide (*travel*) _____ to Paris, I will do my best!

Part B *Directions:* Complete the sentences with the correct prepositions.

1. Do you believe _____ ghosts?

2. You can reach the top shelf _____ using the stool that's next to the table.

3. In some areas of my country, it's cheaper to travel _____ train than _____ bus.

4. Mrs. Miller always looks forward _____ seeing her grandchildren.

5. We paid our electric bill _____ credit card, but the company has no record of it.

6. My mother-in-law made this baby blanket _____ hand.

7. I'm sure I can quickly repair your pants _____ a needle and thread.

8. New drivers are excited _____ driving, but they are sometimes afraid _____ getting in an accident.

Part C *Directions:* Complete the sentences with the correct form of the given adjective and *too* or *enough*.

1. I can't hear the speaker's voice. It's so quiet.

    a. loud   The speaker's voice isn't _____ to hear.

    b. quiet   The speaker's voice is _____ to hear.

2. I let my coffee sit on the table too long. Now I don't want to drink it.

    a. warm   It isn't _____ to drink.

    b. cool   It is _____ to drink.

3. The sun is coming in through the window. It's so bright that we can't see the picture on the TV screen.

    a. bright   It is _____ to see the picture.

    b. dark   It isn't _____ to see the picture.

Part D *Directions:* Check (✓) the correct sentence in each pair.

1. a. _____ Eating popcorn and watching a movie is a great way to relax in the evening.

   b. _____ To eat popcorn and to watch a movie is a great way to relax in the evening.

2. a. _____ I need to stop at the ATM for getting some money.

   b. _____ I need to stop at the ATM to get some money.

3. a. _____ Are you responsible for cleaning up after the party?

   b. _____ Are you responsible to clean up after the party?

4. a. _____ Is relaxing to walk barefoot in the sand on the beach.

   b. _____ It's relaxing to walk barefoot in the sand on the beach.

5. a. _____ Jeannie asked to leaving work early to take her children to the doctor's office.

   b. _____ Jeannie asked to leave work early to take her children to the doctor's office.

6. a. _____ William took a class to learn how to use photo-editing software.

   b. _____ William took a class for learn how to use photo-editing software.

7. a. _____ Driving in heavy traffic it requires skill and patience.

   b. _____ Driving in heavy traffic requires skill and patience.

8. a. _____ I cleaned the floor with a mop.

   b. _____ I cleaned the floor by a mop.

9. a. _____ It is sometimes scary to visit the doctor for young children.

   b. _____ It is sometimes scary for young children to visit the doctor.

CHAPTER 14 Noun Clauses

QUIZ 1 Identifying Noun Clauses (Chart 14-1)

Directions: Underline the noun clause in each sentence.

Example: All my friends know <u>where I live</u>.

1. I don't know what her name is.
2. Nick doesn't know what time the movie starts.
3. My son wanted to know if I needed help with the dishes.
4. I didn't know that it snowed in April last year.
5. My parents want to know who lives in the big house on the corner.
6. No one knows where Nathan works.
7. My teacher doesn't know whose book this is.
8. James knows that I take the bus to school.
9. Ruth wants to know if I will go to Montreal next summer.
10. I don't know where my keys are.

QUIZ 2 Noun Clauses with Question Words (Chart 14-2)

Directions: Complete Speaker B's responses with noun clauses. Add final punctuation.

SITUATION: *Children's Questions*

Example: A: Why is the sky blue?

    B: I don't know _____*why the sky is blue*_____ .

1. A: Why does the Earth have a moon?

    B: I'm not sure _____.

2. A: Where does the sun go at night?

    B: I really can't say _____.

3. A: When do animals go to bed?

    B: I have no idea _____.

4. A: How does a chameleon change its color?

    B: I wonder _____.

5. A: How tall is a giraffe?

    B: I am not sure _____.

(continued on next page)

6. A: Who planted all the trees in the forest?

   B: I can't tell you _____.

7. A: Whose trees are they?

   B: I don't know _____.

8. A: What color is the ocean?

   B: I'm not sure _____.

9. A: How many fish live in the ocean?

   B: I don't know _____.

10. A: Where is the bottom of the ocean?

    B: I can't say exactly _____.

QUIZ 3 Noun Clauses and Information Questions (Charts 5-2 and 14-2)

Directions: Complete the conversations with the words in parentheses to make noun clauses or information questions.

Example: A: Where (*Harry, go*) ___*did Harry go*___ on his vacation last year?

B: I'm not sure where (*he, go*) ___*he went*___. He usually goes camping.

1. A: How much (*an apartment, cost*) _____ in Seattle?

   B: I don't know how much (*rent, be*) _____, but I know Seattle is an expensive place to live.

2. A: Can you tell me when (*the conference, start*) _____?

   B: I'm not sure what (*the date, be*) _____, but we can check online.

3. A: What (*Hoang, have*) _____ to do for homework tonight?

   B: I'm not sure. I haven't asked him how much (*he, have*) _____ to do.

4. A: Why (*that lady, cry*) _____ Mommy?

   B: I don't know why (*she, cry*) _____, but she looks very sad.

5. A: Somebody left a really beautiful velvet jacket in my car. Do you know whose jacket (*it, be*) _____?

   B: Hmmmm ... What color (*the jacket, be*) _____?

   A: It's red.

QUIZ 4 Noun Clauses with *If* and *Whether* (Chart 14-3)

Directions: Complete the conversations with noun clauses. Use *if* to introduce the noun clause.

Example: A: Are you hungry?

B: Not really, I had a big lunch.

A: Well, please tell me ___*if you are*___ hungry later. I can make us a snack.

1. A: Is Norma a lawyer?

   B: I don't know _____ a lawyer, but she works in a law office.

2. A: Do you have a phone I can use?

   B: What did you say?

   A: I want to know _____ a phone I can use.

3. A: Does Oscar want french fries with his hamburger?

   B: Why are you asking me? I don't know _____ french fries with his hamburger. You need to ask him.

4. A: Are you going to the movies this weekend?

   B: I haven't decided _____ to the movies this weekend or not.

5. A: Will you have time to go shopping this afternoon?

   B: I'm not sure _____ time to go shopping. I'm pretty busy today.

6. A: Is your brother coming over this weekend?

   B: I'm sorry. What did you say?

   A: I want to know _____ over this weekend.

7. A: Did Sara borrow the car?

   B: I don't know. Why don't you ask your dad _____ the car.

8. A: Does this bus go downtown?

   B: I'm not sure _____ downtown. You should check with the driver.

9. A: Has Jason finished his report yet?

   B: I haven't talked to him. You'll have to ask him _____ his report yet.

10. A: Are Joanna and Max out of town this week?

    B: I don't know _____ out of town this week, but they usually go on vacation at this time of year.

QUIZ 5 Noun Clauses with *That* (Charts 14-4 and 14-5)

Directions: Add the word ***that*** to mark the beginning of a noun clause.

Example: My dad doesn't think ^*that*^ the team will win many games this season.

1. The police are trying to prove the bank manager stole the money.
2. My parents were upset I lost my phone.
3. We're disappointed you didn't believe us.
4. Did I tell you we are moving next month?
5. I heard there's a new Korean restaurant in town.
6. For centuries, people were convinced the Earth was flat.
7. Is it true your dad is a movie script writer?
8. Can you believe it's already June?
9. I'm sure "scissors" starts with "S-C."
10. Carlos was impressed Juan knew so much about chemistry.

QUIZ 6 Substituting *So* or *Not* for a *That*-Clause (Chart 14-6)

A. *Directions:* Complete the conversations with ***so*** or ***not***. Use the information in parentheses to help you.

Example: A: Is it supposed to rain this weekend?

B: I think ___so___. (*It is probably going to rain.*)

1. A: Are you ready for the test?

    B: I believe _____. (*I am ready.*)

2. A: Here's a map. Do you think you can find the street the Browns live on?

    B: I think _____. (*I can find it.*)

3. A: It's started to rain. Do you still want to go swimming?

    B: I guess _____. (*I don't want to go swimming.*)

4. A: You have some money with you. Is it enough?

    B: I suppose _____. (*It is enough money.*)

5. A: Did you forget your wallet?

    B: I hope _____. (*I didn't forget my wallet.*)

(continued on next page)

B. Directions: Restate Speaker B's response by using a ***that*-clause**.

Example: A: Is there a fire alarm in this building?

B: I hope so.

_____I hope that there is a fire alarm in this building._____

1. A: Is Sergei well enough to return to work?

    B: I think so.

    _____

2. A: Is James telling the truth?

    B: I don't believe so.

    _____

3. A: Have you learned the names of all the students yet?

    B: I hope so.

    _____

4. A: Do you want another cup of coffee?

    B: I don't think so.

    _____

5. A: Is Amy going to visit us next month?

    B: I believe so.

    _____

QUIZ 7 Quoted Speech   (Chart 14-7)

Directions: Add quotation marks, capital letters, and punctuation to show quoted speech.

Example: Josef said, "popcorn is my favorite snack food."

1. Carmen asked do you have money for parking
2. The doctor said you must stop smoking immediately
3. There's a mouse in the house my mother yelled
4. The policeman said may I see your driver's license
5. Mary asked did you get the message I left for you
6. The Johnsons said we have to leave now we have another party to attend tonight
7. Our teacher asked who knows the answer who would like to write it on the board
8. My mother said I won't be home until 7:00 tonight
9. Do you speak Russian asked Natasha
10. Miguel asked are you tired from the walk around Green Lake

QUIZ 8 Quoted Speech (Chart 14-7)

Directions: Punctuate the quoted speech in the conversation. The first example is done for you.

SITUATION: *A Conversation Between a Father and Daughter*

1. "I don't like spiders," my daughter said.
2. Why not I asked.
3. They're quite ugly she replied.
4. Well, they might look unpleasant I said. They're not as beautiful as butterflies, but they're good to have around. They eat ants and flies that you don't want to have in your house I said. Try to think of them as a gift from nature.
5. Wow she said. I didn't know spiders were so helpful.
6. You learn something new every day I said.
7. I guess you're right, Dad she answered.

QUIZ 9 Quoted Speech vs. Reported Speech (Chart 14-8)

Directions: Complete the sentences with the correct pronouns or possessive adjectives for reported speech.

Example: Mrs. Adams said, "My granddaughters are going to visit me in the summer."

Mrs. Adams said that ___her___ granddaughters were going to visit ___her___ in the summer.

1. Mrs. Diaz said, "My secretary is on vacation this week."

   Mrs. Diaz said that _____ secretary was on vacation this week.

2. Laura said, "The book that you lent me was really good."

   Laura said that the book that _____ had lent _____ was really good.

3. My husband said, "Our children want us to take them to a movie tonight."

   My husband said that _____ children wanted _____ to take _____ to a movie tonight.

4. Joan said, "I can come to your house tonight and help you and your brother with the science project."

   Joan said that _____ could come over to _____ house tonight and help _____ and _____ brother with the science project.

(continued on next page)

5. Mr. Owens said, "I want my children to be independent and think for themselves."

Mr. Owens said that _____ wanted _____ children to be independent and think for _____ .

6. The woman at the bakery said, "Our customers bought all our fresh bread this morning. We don't have any more bread."

The woman at the bakery said that _____ customers had bought all _____ fresh bread this morning. _____ didn't have any more bread.

7. Mrs. Larson, the school principal, told the parents, "The teachers and I have high standards for your children, and we will support their efforts to be successful."

Mrs. Larson, the school principal, told the parents that _____ had high standards for _____ children, and _____ would support _____ efforts to be successful.

QUIZ 10 Verb Forms in Reported Speech (Chart 14-9)

Directions: Complete the reported speech sentences. Use formal verb forms.

Example: Teresa said, "I'm looking forward to my trip to Washington, D.C."

Teresa said that she __was looking__ forward to her trip to Washington, D.C.

1. The teacher said, "The test will be on Friday."

The teacher said that the test _____ on Friday.

2. The mail carrier said, "There was no mail delivery on Monday because of the holiday."

The mail carrier said that there _____ no mail delivery on Monday because of the holiday.

3. Marie said, "My English teacher has the flu."

Marie said that her English teacher _____ the flu.

4. Junko said, "I don't understand the problem."

Junko said that she _____ the problem.

5. The director said, "The actors haven't learned their lines yet."

The director said that the actors _____ their lines yet.

6. Pedro said, "The movie is going to start in ten minutes. We need to leave."

Pedro said that the movie _____ in ten minutes, and that they _____ to leave.

(continued on next page)

7. The weather reporter said, "There has been heavy snow in the mountains for the past few days."

   The weather reporter said that there _____ heavy snow in the mountains for the past few days.

8. The Smiths said, "We can feed your cats while you are away."

   The Smiths said they _____ our cats while we _____ away.

QUIZ 11 Reported Speech   (Chart 14-8 and 14-9)

Directions: Change the quoted speech to reported speech. Use formal verb forms.

Example: My grandfather said, "I have had a wonderful life."

   My grandfather said (that) he had had a wonderful life.

1. Abdul said, "I will be 25 on my next birthday."

2. My parents said, "We enjoyed our trip to Costa Rica."

3. My friends said, "We want to have a birthday party for you."

4. Suzanne said, "I have lived in Italy for 20 years."

5. The boy said, "The dog took my ball. He isn't coming back."

6. Emily said, "I'm going to retire in a few years. My husband and I are planning to travel."

7. My husband said, "I can pick up the kids after school. You don't need to worry about it."

QUIZ 12 Say, Tell, and Ask (Chart 14-10)

Directions: Complete the sentences with the correct reporting word. Use **said**, **told**, or **asked**.

Example: Ben ___told___ me he could take me to the train station.

1. The teacher _____ me if I had done my homework. I _____ her I had finished it last night.

2. The manager _____ that all employees would get extra pay for the holiday. The employees were pleased and _____ the manager that they were grateful.

3. Scott _____ his parents that he had rented an apartment. His parents _____ they were pleased that he had found one.

4. The chef on the children's TV baking show _____ to use one cup of sugar, but the children heard it wrong and used two. They asked their mom what to do, and she _____ them to double the recipe.

5. The security guard _____ that everyone needed an ID card to enter the building. I _____ him that I had left mine at home. I _____ him if I could show him my driver's license instead. He _____, "No." He _____ me I had to go home and get my ID card if I wanted to go in. I _____ him I would be late for work. He _____ that he was sorry, but he couldn't change the rules.

QUIZ 13 Chapter Review

Directions: Correct the errors.

Example: My sister ~~said~~ *told* me that she was going to Los Angeles for a week.

1. I don't know the doctor can see you tomorrow or not.
2. Please tell me what did they do.
3. Do you know whose coat on the chair is?
4. Do you know if or not the bus has come?
5. We'd like to know if the subway stop here?
6. I hope so that I can come with you tonight.
7. Leila told that she wasn't home last night.
8. I'm sorry what we have to cancel our plans.
9. The dentist said Your teeth look very healthy. You are taking good care of them.
10. The teacher isn't sure whether Liz wants help or no.

CHAPTER 14 – TEST 1

Part A *Directions:* Complete Speaker B's responses with noun clauses.

1. A: Mom, where's the milk?

    B: I don't know _____.

2. A: Mr. Barrett, what time will we be finished?

    B: I'm not sure _____.

3. A: I'm hungry. Are there any eggs left?

    B: I don't know _____.

4. A: Marcos, whose homework is that?

    B: I don't know _____.

5. A: Is someone knocking at the door?

    B: I wonder _____.

6. A: I just heard the phone ring. Who called?

    B: I'll find out _____.

7. A: Did Kwon finish his biology lab work?

    B: Let's ask Maiko. She probably knows _____.

8. A: What is the weather supposed to be like tomorrow?

    B: I haven't checked the forecast. I don't know _____.

9. A: Has anyone met the new coach yet?

    B: I haven't. I'm not sure _____.

10. A: Do we have to type this assignment?

    B: I'll ask the teacher _____.

Part B *Directions:* Punctuate the quoted speech in the text.

SITUATION: *A Conversation Between a Mother and Daughter*

1. Mom came in to wake me up. What time is it I asked her.

2. It's time to get up she replied.

3. But it's not a school day I said. Please let me sleep in I begged.

4. You can't sleep in today she said. It's a special day.

5. What special day I asked.

6. It's your birthday she said.

7. Oh my gosh! I forgot. I have to get up right away I said. I have so many things I want to do today.

Part C *Directions:* Change the quoted speech to reported speech. Use formal verb forms.

1. Julia said, "The cookies are ready."

2. The librarian said, "The library is going to close early today."

3. John asked, "How far away is the airport?"

4. The fire chief said, "It took a long time to put the fire out."

5. The clerk asked me, "Do you want to pay with cash or a card?"

6. Marika said, "The flight will arrive in ten minutes."

7. The teacher said, "The test is going to be on Friday."

8. The students replied, "We don't want a test."

9. The manager said, "You need to put your cell phones on silent in the office."

10. Joan asked, "Have you ever watched a movie in Swedish?"

Part D *Directions:* Correct the errors.

1. My friends understand what do I like.
2. I'd like to know does this computer work?
3. The teacher told that he would be at a meeting tomorrow.
4. I want to know why did they come.
5. Is a fact that exercise makes us healthier.
6. Do you know if Rick live here.
7. I'm sure, that we will have a good time together.
8. A classmate asked me "Where I live."
9. Do you know whose are these keys?
10. I'm not sure that if he wants to come or not.

CHAPTER 14 – TEST 2

Part A *Directions:* Complete Speaker B's responses with noun clauses.

1. A: What's the date today?

   B: I'm not sure _____.

2. A: Do you know what year Aaron was born?

   B: I can't remember _____.

3. A: Did Dimitri and Irina get engaged last weekend?

   B: I have no idea _____.

4. A: Water is leaking under the sink. Has anyone called a plumber yet?

   B: I'll find out _____.

5. A: How many people knew about the problem?

   B: I'll see _____.

6. A: Does the weather here change much at different times of year?

   B: I don't know _____.

7. A: I know Paula is unhappy. Did she leave the company?

   B: I haven't heard _____.

8. A: Who will be the new manager?

   B: We don't know _____.

9. A: Whose car are we taking to the mall?

   B: We haven't decided _____.

10. A: Does the copy machine work?

    B: I'll see _____.

Part B *Directions:* Punctuate the quoted speech in the text.

SITUATION: *A Conversation Between a Teacher and a Student*

1. What do you want to do after you finish school my teacher asked.

2. I'm not sure I said. I'd like to have a job that is interesting and pays well.

3. Everyone would like that said my teacher. Is there a specific area you see yourself working in?

4. Yes I replied. I love working with animals. Maybe I could be a veterinarian.

5. One way to find out is to work with animals first said my teacher. Volunteer at an animal shelter or zoo. See how you like it.

6. I told him I like that suggestion. Thanks!

Part C *Directions:* Change the quoted speech to reported speech. Use formal verb forms.

1. The reporter asked me, "Do you have time to answer a few questions?"

2. Yolanda said, "I cleaned my apartment and did my laundry."

3. My friend said, "The bus will be late."

4. Joe asked, "Who took my car?"

5. The manager said, "We have decided to move our offices to a new location."

6. Brad asked, "When will the book be published?"

7. Shirley said, "I can fix that for you."

8. My parents said, "We were happy to hear about your new job."

9. The doctor asked me, "Have you been taking your medicine?"

10. The dancers said, "We're ready for our show."

Part D *Directions:* Correct the errors.

1. Can you tell me whose coat is this?

2. The doctor said, your daughter just has a bad cold. It's nothing serious.

3. My friends asked me when will you get married.

4. Did my mom ask you if or not you could come to our party?

5. Hamid asked why did I always come late?

6. Mr. Hill told to me that he was feeling ill.

7. Could you tell me where Fred work in the evenings?

8. I think so that you will enjoy being on the soccer team.

9. I know, that this will be a good opportunity for us.

10. Professor Thomas told us he will be absent yesterday.

MIDTERM EXAM 1 — Chapters 1-7

Directions: Circle the correct completions.

Example: Maria and Tony ____ famous characters from *West Side Story*.

    a. is   c. are being

   (b.) are   d. has been

1. The Duncans and their son ____ dinner together on Sundays.

   a. eats usually   c. eat usually

   b. usually eats   d. usually eat

2. You ____ have a passport to travel to a foreign country.

   a. must   c. may

   b. can   d. could

3. Almost everyone in our neighborhood ____ a dog or a cat.

   a. have   c. has

   b. does have   d. is having

4. ____ is it from Barcelona to Madrid?

   a. How far   c. How long

   b. How often   d. How much time

5. You have a good job, ____ you?

   a. are   c. do

   b. aren't   d. don't

6. Some people like chocolate ice cream, but ____ prefer vanilla.

   a. other   c. the other

   b. others   d. another

7. I have been working at the same company ____ .

   a. next year   c. since 2001

   b. until March   d. from now

8. Right now Marcy ____ the piano. She enjoys music.

   a. is playing   c. play

   b. plays   d. does play

9. _____ Ms. Simmons take the bus to work?

   a. Is   c. Does

   b. Are   d. Do

10. Stephan _____ coffee. He prefers tea.

    a. don't drink   c. aren't drink

    b. doesn't drink   d. isn't drink

11. Your daughter really enjoyed the ballet, _____ she?

    a. didn't   c. wasn't

    b. won't   d. can't

12. Jerry and Jennifer _____ go to Macedonia next year. They aren't sure yet.

    a. must   c. should

    b. maybe   d. might

13. I got a text message while I _____, so I couldn't answer it.

    a. drove   c. am driving

    b. drives   d. was driving

14. We have to pick up Travis in a few minutes. I have _____ address.

    a. him   c. he

    b. he's   d. his

15. Paul was slicing a tomato when he cut _____ and had to go to the hospital.

    a. him   c. his

    b. himself   d. he

16. My grandparents _____ to Venice many times. They go there every year.

    a. are going   c. have been

    b. has been   d. was going

17. The baby _____ won't sleep much this afternoon. She had a long nap this morning.

    a. may   c. probably

    b. maybe   d. may be

18. Tony is moving to a new apartment tomorrow. I ____ help him move.

    a. going to
    b. am going to
    c. am
    d. will be

19. ____ does the bank open today?

    a. Where
    b. What time
    c. What
    d. Who

20. Katya and her roommate ____ together at the library last night.

    a. study
    b. was studying
    c. studied
    d. will study

21. I ____ my homework by 10:00, so I went to bed.

    a. finish
    b. was finished
    c. had finished
    d. had been finished

22. **A:** Are you a good cook?

    B: Yes, I ____. I like to cook.

    a. am
    b. do
    c. is
    d. did

23. I have a cold. I ____ terrible yesterday, but I'm feeling better today.

    a. feel
    b. felt
    c. feels
    d. fell

24. ____ Andrew drive or take the train to London last weekend?

    a. Does
    b. Was
    c. Did
    d. Were

25. I saw a ____ at the mall yesterday, but it was too expensive.

    a. beautifuls bag
    b. bag beautiful
    c. beautiful bag
    d. bag beautifuls

26. My parents ____ an old truck, but now they have a new sports car.

    a. used to drive
    b. use to drive
    c. used to driving
    b. use to driving

27. ____ after you finish checking your email?

    a. What are you do
    b. What you going to
    c. What are you going to do
    d. When are you going to

28. My brother ____ in ghosts, but I don't.

    a. belief
    c. believes
    b. believe
    d. is believing

29. In a few months, we ____ to Oregon.

    a. going to go
    c. are going
    b. will going
    d. going

30. I hate my new schedule! I ____ getting up at 6:30 every morning.

    a. don't used to
    c. didn't used to
    b. am not used to
    d. wasn't use to

31. When ____ clothes are dry, I will fold them and put them away.

    a. Marias
    c. Marias'
    b. Maria's
    d. Maria

32. ____ you please help me with my algebra? I don't understand this problem.

    a. Should
    c. Must
    b. May
    d. Would

33. ____ Ms. Carlson going to teach a grammar class next semester?

    a. Is
    c. Does
    b. Are
    d. Will

34. It was really cold yesterday, so we had ____ for dinner.

    a. beans soup
    c. bean soup
    b. soup bean
    d. soups bean

35. Freddy's soccer team ____ play in a tournament during the summer.

    a. maybe
    c. may be
    b. may
    d. be

36. As soon as my brother ____, he'll buy us some ice cream.

    a. comes
    c. came
    b. will come
    d. has come

37. My plane ____ at 10:50 Friday night.

    a. leaving
    c. has left
    b. has left
    d. leaves

38. What _____ since I saw you last summer?

   a. are you doing
   b. you are doing
   c. have you been doing
   d. you have been doing

39. Mr. Robbins has taught high school French _____.

   a. since many years
   b. for 15 years
   c. three years ago
   d. last year

40. We _____ Krista since she had her baby two months ago.

   a. hasn't seen
   b. haven't seen
   c. didn't see
   d. wasn't seen

41. Students _____ cheat on their exams or they will get in trouble.

   a. don't have to
   b. didn't have to
   c. had better not
   d. might not

42. When we arrived at the theater, the movie _____.

   a. already started
   b. had already started
   c. has already started
   d. is started

43. _____ lives in the house at the end of the street?

   a. Where
   b. Whom
   c. Who
   d. Whose

44. It's been a long time since you've seen your parents, _____ it?

   a. wasn't
   b. haven't
   c. hasn't
   d. isn't

45. Leila's job interview is at 1:00 P.M. _____ Thursday.

   a. on
   b. to
   c. in
   d. at

46. People around the world _____ the internet to get the news.

   a. uses
   b. use
   c. is using
   d. has used

47. One of Beth's favorite sports is tennis. _____ is swimming.

   a. Other
   b. The others
   c. Others
   d. Another

48. When Mr. Barry gave his presentation, people in the back of the room ____ hear him.

    a. couldn't   c. mustn't

    b. shouldn't   d. might not

49. ____ the student with the highest grade in the class?

    a. Who   c. Who's

    b. Whose   d. Whom

50. It has been snowing for two days. The skiing ____ be great.

    a. have to   c. can

    b. had better   d. must

MIDTERM EXAM 2 Chapters 1–7

Part A *Directions:* Complete the sentences with an appropriate form of the words in parentheses. More than one answer may be correct.

Example: Helena didn't want anyone to find her diary, so she (*hide*) _____hid_____ it in her closet.

1. When I went shopping yesterday, I (*buy*) _____ some light bulbs and a broom.

2. Every year, my whole family (*get*) _____ together to celebrate my grandfather's birthday.

3. Saya (*fall*) _____ as she was getting on the bus. She (*be*) _____ so embarrassed.

4. Next September, my daughter (*start*) _____ high school. She is growing up!

5. I (*be, never*) _____ to Turkey, but I think it would be an interesting place to visit.

6. By the time the chemistry lecture ended, Faisal (*take*) _____ six pages of notes.

7. Right now, the children (*draw*) _____ pictures of themselves in art class.

8. Juan (*have, not*) _____ a car. He (*take*) _____ the train to work every day.

9. Emma and Hank (*move*) _____ into their new house next weekend.

10. Vinh used to have a motorcycle, but he (*sell*) _____ it because he needed money.

11. Mr. Smetko (*work*) _____ for the same company for more than 29 years. Next year, he (*retire*) _____.

12. Anna didn't want anyone to see her bad grade, so she (*throw*) _____ her paper away.

13. Nathaniel was about to send a text to his girlfriend when he noticed that he (*get, already*) _____ a text from her.

14. After I finish taking these pictures, I (*send*) _____ them to my parents. They (*love*) _____ seeing pictures of me having fun with my friends.

15. Many people (*cheer*) _____ when the marathon runners crossed the finish line. The runners (*smile*) _____.

Part B *Directions:* Complete the sentences with *can, could, may, might, would, should, had better,* or *must*. More than one answer may be correct.

Example: I used to get up at noon, but now I _____must_____ be at work by 8:00 A.M.

1. People _____ not talk on their cell phones while driving.
2. _____ I please have a hamburger and french fries?
3. It's raining again. It _____ rain all day. I'm glad I brought my umbrella.
4. Hans is late! He _____ hurry up or he'll miss the exam.
5. Nancy just got a big promotion at work. She _____ feel great!
6. When I was young, I _____ not speak German, but now I speak it well.
7. _____ you please turn down your music? It's too loud.
8. I think you _____ try the chocolate chip cookies at Branca's Bakery. They're delicious!
9. I enjoy listening to music, but I _____ not sing at all. I've got a terrible voice.
10. Children _____ be respectful when talking with adults.

Part C *Directions:* Complete the questions and answers with the correct words. More than one answer may be possible.

Examples: A: _____Does_____ this road go through the center of town?

B: No, _____it doesn't_____.

1. A: Excuse me. _____ there a good restaurant near here?

   B: Well, _____ kind of food do you want?

   A: Italian, maybe. _____ you know of any Italian restaurants in this area?

   B: Yes, _____. There is one on Fremont Street that's pretty good.

2. A: _____ wireless speaker is that?

   B: It's mine. _____ you want to use it?

   A: Yes, _____. _____ long will the battery last?

   B: It should play for at least 18 hours. _____ do you need to know?

   A: Because I want to take it to a party. _____ that OK?

   B: Sure, but please recharge it when you are finished.

3. A: You look tired. _____ you sleep well last night?

   B: No, _____ .

   A: That's too bad. _____ did you go to bed?

   B: At 1:00 A.M.

   A: Wow! _____ did you stay up so late?

   B: I had too much homework to do. _____ homework did you have?

   A: Not much. I finished it and went to bed early.

Part D *Directions:* Correct the errors.

Example: Most ~~leaf~~ *leaves* are green.

1. Tom last name is Miller.
2. Tomato are good for us. They have lots of vitamin C.
3. They would rather eat Chinese food, don't they?
4. All of the actors names are listed on page six of your program.
5. Do you want to go shopping on Saturday? I need a new shoes.
6. Mr. Tobias works in his flowers garden every day.
7. My brothers started they're own gardening business last year.
8. Some children prefer to play by theirselves rather than with other kids.
9. There are two new students in class. One is from Libya, and another is from Romania.
10. Jason forgot him math book on the bus. I hope he can get it tomorrow.

FINAL EXAM 1 — Chapters 1–14

Directions: Circle the correct completions.

Example: Diana ___ to her cousin's birthday party.

   a. invited
   (c.) was invited
   b. had invited
   d. invite

1. Can you please take this package to ___ post office?

   a. Ø
   c. an
   b. the
   d. some

2. My kids must clean their rooms ___ do laundry on Saturdays.

   a. but
   c. and
   b. because
   d. so

3. People should always put on ___ sunscreen lotion to protect their skin.

   a. the
   c. an
   b. a few
   d. Ø

4. Ted ___ horror movies, and neither does his girlfriend.

   a. doesn't like
   c. don't like
   b. isn't like
   d. likes

5. *Too*, *two*, and *to* have ___ pronunciation but different meanings.

   a. a same
   c. the same
   b. like
   d. alike

6. Max has a job painting houses. He ___ on a tall ladder.

   a. use to stand
   c. is used to standing
   b. uses to stand
   d. was used to standing

7. Babies ___ by everyone.

   a. was loved
   c. loved
   b. are loved
   d. have loved

8. San Francisco has many tourist attractions, and ___ Los Angeles.

   a. does too
   c. also does
   b. so do
   d. so does

9. I'm going to go to bed early ___ I'm not feeling well.

   a. even though
   c. because
   b. so
   d. but

10. Willie is really good ____ computers. He can solve any computer problem!

    a. on fixing
    b. to fixing
    c. fixing
    d. at fixing

11. My father asked me if I ____ money for gas.

    a. needed
    b. needs
    c. needing
    d. did need

12. My new cell phone was not ____ my old one.

    a. expensiver than
    b. more expensive as
    c. as expensive as
    d. more expensive

13. My sister has three kids, ____ she is very busy.

    a. because
    b. so
    c. or
    d. although

14. Abdul is thinking about ____ medicine, but he has to take the medical school exam first.

    a. study
    b. studying
    c. to study
    d. going study

15. You forgot to bring me the money you owe me, ____?

    a. don't I
    b. aren't you
    c. didn't I
    d. didn't you

16. The woman ____ I share my apartment with is a doctor.

    a. which
    b. whom
    c. what
    d. whose

17. Rudy didn't have to go to the dentist, but his brother ____.

    a. does
    b. was
    c. would
    d. did

18. Hiromi is looking forward ____ to Hawaii on her honeymoon.

    a. going
    b. to go
    c. to going
    d. on going

19. The loan officer ____ helped us with our loan was knowledgeable and friendly.

    a. Ø
    b. who
    c. she
    d. whom

20. My daughter understands Italian much better than I ____.

    a. do
    b. am
    c. will
    d. did

21. Fiona's shoe broke while she was dancing. She was really ____.

    a. embarrassed
    b. embarrass
    c. embarrassing
    d. embarrasses

22. Several of my classmates are funny, but Alexis is ____ of all.

    a. funniest
    b. the funniest
    c. most funny
    d. the most funny

23. German, French, and Italian ____ in different areas of Switzerland.

    a. are spoken
    b. is spoken
    c. speak
    d. speaks

24. The little boy didn't remember where ____.

    a. he did live
    b. lived he
    c. did he live
    d. he lived

25. Fresh fruit and vegetables should ____ every day as part of a healthy diet.

    a. eat
    b. to eat
    c. be eaten
    d. eating

26. Naomi ____ her upcoming trip to Australia.

    a. excited for
    b. is excited
    c. excites about
    d. is excited about

27. When I lived in London, I ____ English every day. My English got a lot better.

    a. use to speak
    b. used to speak
    c. am used to speaking
    d. used to speaking

28. I'm sorry, but I can't go out tonight. I have to study for ____ exam.

    a. some
    b. Ø
    c. a
    d. an

29. ____ in Maha's ring is 24 karats.

    a. Gold
    b. The gold
    c. Golds
    d. A gold

30. How ____ did George spend on his new suit?

    a. many money
    b. many monies
    c. much money
    d. much monies

31. My parents ____ me I had to be home by midnight.

    a. said
    b. asked
    c. told
    d. telled

32. Anne lost her necklace. It was ____ necklace that her husband had given her on their first anniversary.

    a. Ø
    b. an
    c. the
    d. some

33. I wonder ____ this weekend. I want to go on a picnic.

    a. if it rains
    b. if it will rain
    c. if or not it will rain
    d. will it rain

34. Do you know the couple ____ son won the speech contest?

    a. their
    b. which
    c. whose
    d. who

35. They ____ in Chicago for ten years, and they really like it.

    a. lives
    b. are living
    c. have lived
    d. live

36. I had ____ in my pocket, so I gave them to the street musicians.

    a. a few coins
    b. few coins
    c. a little coins
    d. little coins

37. Do you know the boy ____ wearing the striped shirt?

    a. who's
    b. he's
    c. whose
    d. what's

38. Scientists want to do ____ research on whether life can exist on Mars.

    a. farther
    b. further
    c. farthest
    d. furthest

39. My friend Shoko goes ____ every Saturday.

    a. shop
    b. shops
    c. to shop
    d. shopping

40. Whenever I buy plane tickets, I pay ____ a credit card.

    a. by
    b. for
    c. with
    d. on

41. Hiro studied hard ___ high grades in all his classes.

    a. for getting
    b. to get
    c. on getting
    d. for get

42. The chicken is ___. I can't eat it.

    a. spicy enough
    b. to spicy
    c. too spicy
    d. spicier

43. My classes this semester aren't ___ at all. All of my professors are really good!

    a. bored
    b. bore
    c. boring
    d. bores

44. What time ___?

    a. does the movie start
    b. does start the movie
    c. the movie starts
    d. starts the movie

45. Getting my driver's license was ___ easier than I expected.

    a. more
    b. much
    c. many
    d. less

46. **A:** Have you ever visited Istanbul?

    B: No, I ___.

    a. didn't
    b. am not
    c. don't
    d. haven't

47. Emanuel can't decide ___ stay here or go back to France.

    a. should he
    b. if should he
    c. whether he should
    d. whether or not

48. Is it true ___ we only have three weeks of school left?

    a. what
    b. do
    c. if
    d. that

49. Our teacher said that we ___ to turn in our homework at the beginning of class.

    a. have
    b. had
    c. have had
    d. are having

50. ___ flowers are these? They're beautiful!

    a. Who
    b. Whom
    c. Who's
    d. Whose

FINAL EXAM 2 Chapters 1–14

Part A *Directions:* Combine the sentences with the words in parentheses. Add any necessary punctuation. More than one answer may be possible.

Example: Ellis is avoiding sugar. Ellis didn't order any dessert. (*because*)

    Because Ellis is avoiding sugar, he didn't order any dessert.

1. Mark wants to go to Alaska next summer. His wife would rather go to California. (*but*)

2. Mr. Meecham went to a meeting with a new client. He wore his best suit. (*when*)

3. This radio station plays lots of classic rock music. I enjoy listening to it. (*so*)

4. Chris is a great guitar player. Monica is a fantastic pianist. (*and*)

5. Nan and Rose both enjoy playing tennis. They don't play together very often. (*although*)

6. We finished dinner. We had strawberry shortcake for dessert. (*as soon as*)

Part B *Directions:* Complete the sentences with a word or phrase from the box. Use each word or phrase only once. The first one is done for you.

| whether or not | how long | what time | if | who | which | that |

1. You need to finish this report by tomorrow. I don't care ____*if*____ it takes you all night.

2. Do you know _____ train goes to Sarajevo?

3. The students _____ acted in the play did a wonderful job!

4. I'm not sure _____ the show lasts—probably about two hours.

5. It's true _____ spiders are helpful insects in the garden.

6. I don't know _____ James got a high grade on his research paper. I'm sure he learned a lot from writing it.

7. I forgot _____ my appointment at the dentist was, so I was 15 minutes late.

Part C *Directions:* Add quotation marks, capital letters, and the correct punctuation to show quoted speech.

Example: "What is the name of the book you're looking for?" the clerk asked.

1. i'm so sorry to hear that your father is ill said margaret

2. john asked do you want to have toast or cereal for breakfast

3. could i please have a glass of water mary asked i'm really thirsty

Part D *Directions:* Change the quoted speech to reported speech. Use formal verb forms.

Example: Dora said, "I hope to travel around the world."

      Dora said that she hoped to travel around the world.

1. Derek said, "Social networking websites are a great way to keep in touch."

2. Heidi said, "I have always dreamed of being an actress."

3. John asked, "Which video game do you like the best?"

4. Anita said, "I will go to college next year."

5. My friends asked, "Do you want to go to the coffee shop with us?"

Part E *Directions:* Correct the errors.

Example: Let's go to Julia's Cafe. Their food is ~~good~~ better than the food at Mel's Diner.

1. Adam is taller as his brothers.

2. I'll can help you in just a few minutes.

3. Playing the clarinet is not as difficulty as playing the oboe.

4. That is the funny joke I've heard in a long time.

5. My sisters like country-western music much better than I am.

6. If you don't fix the leak in the roof, it will cause farther water damage.

7. My brother and I are twins, but I didn't see him since 2010.

8. Laptop computers are usually convenienter than desktop computers.

9. How many apples you bought?

10. My classmate has as same name as me. We are both named William.

Part F *Directions:* Complete the sentences with *a*, *an*, *the*, or *Ø*.

Example: Will you please turn off ___the___ light?

1. I saw _____ really funny play last night.
2. _____ books that we borrowed from the library must be returned by Friday.
3. _____ alligator can weigh over 1,000 pounds.
4. It's not good for _____ children to spend too much time online.
5. Yesterday, we went swimming near our house. _____ water was so cold!

Part G *Directions:* Circle the correct completions.

1. **How far / How long** is it from Paris to Marseille?
2. Students in Advanced English have a lot of **homework / homeworks**.
3. Ken can't eat **the / Ø** bananas. He's allergic to them.
4. How **much / many** people will be at the meeting?
5. Patrick needs **a few / a little** more time to prepare dinner. We'll eat in about 15 minutes.
6. Kelly really likes **a / Ø** coffee. It's her favorite beverage.
7. Thousands of blogs **publish / are published** on the internet every day.
8. There are many islands in **the / Ø** Pacific Ocean.
9. Anthropologists are interested in **to learn / learning** about ancient cultures.
10. When I was in Sicily, I went to the top of **the / Ø** Mount Etna.
11. Alyssa invited **several / much** friends to her 16th birthday party.
12. I plan **to go / going** to the supermarket after I finish writing this email.
13. Scientists **have been studying / have been studied** the HIV/AIDS virus since the 1980s.
14. This umbrella isn't **my / mine.** It belongs to Sue.
15. This laptop **gave / was given** to me by my parents.

ANSWER KEY

CHAPTER 1

Quiz 1, p. 1
walk, are walking
walks, is walking
walk, are walking
walk, are walking
walk, are walking

Quiz 2, p. 1
1. eats
2. does not / doesn't eat
3. enjoys
4. walk
5. take
6. drive
7. begins
8. rings
9. come
10. do not / don't have

Quiz 3, p. 2
1. are playing
2. are swimming, (are) diving
3. is making
4. is throwing, is catching
5. are having
6. is looking
7. is flying
8. is getting

Quiz 4, p. 2
A.
1. Does she
2. Does she
3. Is she
4. Does she
5. Is she

B.
1. do not / don't call
2. is not / isn't studying
3. do not / don't remember
4. are not / aren't taking
5. does not / doesn't live

Quiz 5, p. 3
2. noun, plural
3. noun, plural
4. verb, singular
5. noun, plural
6. verb, singular
7. noun, plural
8. verb, singular
9. verb, singular
10. verb, singular
11. noun, plural

Quiz 6, p. 4
A. Suggested answers:
1. is often
2. never pay
3. sometimes forgets
4. always go
5. rarely see

B.
1. Does Alex ever go bowling?
2. We seldom go to the theater more than once a month.
3. Abdul is usually hungry at dinnertime. OR Usually Abdul is hungry at dinnertime.
4. I occasionally stay out past midnight on weekends. OR Occasionally I stay out past midnight on weekends.
5. Lee doesn't always do his homework.

Quiz 7, p. 5
1. is relaxing, is listening
2. comes, arrives
3. is ringing, is calling
4. are cleaning, have
5. isn't working, looks

Quiz 8, p. 5
2. am looking
3. don't remember
4. Are you having
5. am thinking
6. belongs
7. doesn't have
8. think
9. looks
10. are practicing, need

Quiz 9, p. 6
1. is answering, does
2. speaks, is speaking, want
3. is flying
4. rides, is working, is not riding
5. Do you like

Quiz 10, p. 6
A: Do your neighbors have
B: they do, have
A: Is it
B: it is, likes
A: Are they walking
B: they are, loves
A: Do you want
B: I don't

Quiz 11, p. 7
1. go, is
2. looks, Is, is, am trying
3. are having
4. Do, am leaving, don't
5. believe, own, work
6. are you doing, am sending, Do you send, don't have
7. snows, become

Quiz 12, p. 8
1. usually begins, ends
2. are singing
3. is crying, is always
4. Are you cooking, am
5. catches, walks
6. Do you and Sam want, don't
7. are painting
8. is, don't know, is

Quiz 13, p. 9
1. Rudy **wants** to go skateboarding with his friends.
2. My sister **has** four children: two girls and two boys. They are so cute!
3. Stephanie and Sarah **meet** at the coffee shop every Thursday night.
4. **Is it raining** right now?
5. I like to play tennis, but I **am** not a very good tennis player.
6. B: No, I **don't**.
7. Nora doesn't **have** time to watch TV. She has too much homework.
8. Look out! That car **is going** too fast!
9. A: Yuko, what **are** you doing?
10. I **practice** English in class every day.

TEST 1, p. 10

Part A
1. is coming, want
2. don't understand, lets
3. are watching
4. doesn't belong
5. need, is

Part B
1. is getting, is coming
2. practices
3. goes, always plays, are kicking
4. works, often teaches, like, is
5. has, picks, is not, repair, are working

Part C
1. do not / don't like
2. does not / doesn't want
3. is not / isn't
4. does not / doesn't see
5. am not / 'm not

Part D
1. The teacher **never yells** at her students. She is very patient.
2. What time **do** you leave school every day?
3. The Smiths **do not / don't** have a car. They take the bus everywhere.
4. **Does** Jonathan own an apartment or a house?
5. Wait. The sandwiches **are** ready, but not the pizza.
6. Michelle **has** a beautiful engagement ring from her boyfriend.
7. Hey, Kerry. **Do you want** something to drink?

TEST 2, p. 12

Part A
1. cuts, does
2. see, is coming
3. reads, doesn't want
4. goes
5. smells

Part B
1. need
2. plays
3. is, are working, are writing, is correcting
4. own, wake, feeds, takes, have, leaves, catches, are staying, enjoying

Part C
1. do not / don't have
2. is not / isn't
3. do not / don't enjoy
4. does not / doesn't taste
5. are not / aren't going

Part D
1. **Do** you always **go** to school by bus?
2. I **don't** like movies with sad endings.
3. Oh no, look! A rat **is playing** in the garbage can.
4. **Does** Maria work on Saturdays?
5. The books **are** on sale, but not the magazines.
6. Mr. Green is almost 80, but he **doesn't** want to live with his children.
7. Ted eats Mexican food often. He **loves** Mexican food.

CHAPTER 2

Quiz 1, p. 14
1. Andrew worked for eight hours yesterday.
2. Mark and Jan went to bed at 10:00 last night.
3. The alarm on my phone rang at 6:00 yesterday morning.
4. My grandparents visited us last month.
5. I took a nap yesterday afternoon.
6. Dr. Hughes taught medical students last Tuesday evening.
7. Victoria bought coffee at Cory's Coffee Shop yesterday.
8. Mr. Wilson walked to the farmers' Market last Saturday.
9. Anne called her best friend last week.
10. It rained in Seattle last winter.

Quiz 2, p. 15
1. didn't snow
2. didn't travel
3. didn't go
4. wasn't
5. didn't write
6. didn't become
7. didn't drive
8. weren't
9. didn't damage
10. didn't meet

Quiz 3, p. 16
1. Did
2. Were
3. Was
4. Was
5. Did
6. Did
7. Were
8. Did
9. Did
10. Were

Quiz 4, p. 16
1. Did Nina use, Nina didn't use
2. Was the restaurant, The restaurant wasn't
3. Did Mrs. Douglas knock, Mrs. Douglas didn't knock
4. Did Mark take, Mark didn't take
5. Did Julie get, Julie didn't get

Quiz 5, p. 17
1. A: Did Sara eat
   B: she did, ate
2. A: Did you go
   B: I didn't, was
3. A: Did Fred go
   B: he did, hiked
4. A: Did John and Linda travel
   B: they did, spent
5. A: Did you sleep
   B: I didn't, heard, stayed
6. A: Did the monster movie scare
   B: it didn't, didn't scare, made

Quiz 6, p. 18
2. loved, loving
3. helped, helping
4. visited, visiting
5. worried, worrying
6. hopped, hopping
7. played, playing
8. smiled, smiling
9. tied, tying
10. stopped, stopping
11. studied, studying
12. rained, raining
13. preferred, preferring
14. fixed, fixing
15. happened, happening
16. enjoyed, enjoying

Quiz 7, p. 19
2. ran
3. was
4. worked
5. weren't
6. understood
7. tied
8. had
9. tried
10. didn't yell
11. wasn't
12. hurt
13. knew
14. left
15. cleaned
16. didn't have

Quiz 8, p. 20
1. sent
2. took
3. brought, did you buy
4. shook
5. picked
6. asked, didn't hear
7. ordered, was

Quiz 9, p. 20
2. was
3. didn't want
4. didn't tell
5. took
6. woke
7. decided
8. rode
9. got
10. wasn't
11. was
12. shopped
13. bought
14. found
15. ordered
16. walked
17. had
18. sat
19. enjoyed
20. felt
21. was

Quiz 10, p. 21
1. was exercising
2. were visiting
3. were (you) doing
4. wasn't living, was studying
5. were attending
6. was taking, (was) learning
7. Was (he) traveling, was

Quiz 11, p. 21
A.
2. f
3. b
4. e
5. d
6. a

B.
1. Juliette
2. Peter
3. the boys
4. Karen
5. Faisal

Quiz 12, p. 22
Possible answers:
2. Linda met with students while she was working in her office. OR
Linda called her son while she was waiting for the bus. OR
Linda met an old friend while she was sitting at the snack bar.
3. While Janet was working in her office, she checked her email. OR
While Janet was waiting for the bus, she read a novel. OR
While Janet was sitting at the snack bar, she ate a sandwich.
4. Simone answered the phone while she was working in her office. OR
Simone sent a text message while she was waiting for the bus. OR
Simone drank a soda while she was sitting at the snack bar.
5. While Linda was working in her office, she met with students. OR
While Linda was waiting for the bus, she called her son. OR
While Linda was sitting at the snack bar, she met an old friend.
6. Janet checked her email while she was working in her office. OR
Janet read a novel while she was waiting for the bus. OR
Janet ate a sandwich while she was sitting at the snack bar.

Quiz 13, p. 23
1. decided
2. spent, were living, studied / were studying
3. made, won
4. planned
5. ran, was traveling, tried, returned, didn't have
6. had, was blowing / blew, was howling / howled, was trying, became, hid
7. lived, spoke

Quiz 14, p. 24
3. were talking
4. joined
5. chatted
6. decided
7. didn't know
8. was
9. asked
10. wasn't
11. pointed
12. hoped

Quiz 15, p. 24
2. a. 1
   b. 2
3. a. 1
   b. 2
4. a. 1
   b. 2
5. a. 2
   b. 1
6. a. 1
   b. 2
7. a. 2
   b. 1
8. a. 2
   b. 1
9. a. 1
   b. 2
10. a. 2
    b. 1
11. a. 1
    b. 2

Quiz 16, p. 26
1. After Donna got a ticket for speeding, she drove home very slowly.
2. Eric took a shower as soon as he got home.
3. Rick burned his hand while he was cooking dinner.
4. Before Kevin watched a movie, he did his homework.
5. Until my dad got a job in Italy, my family lived in Germany.
6. As soon as I put on my pajamas, I went to bed.
7. Everyone started dancing after the band started to play.
8. Joy's parents drove her to school before she got her driver's license.
9. The students were nervous until the exam began.
10. While Maria and Lucio were exercising, their children were playing a board game.

Quiz 17, p. 27
2. used to swim
3. used to be
4. used to drink
5. used to work
6. used to have
7. used to wake up
8. used to live
9. used to ski
10. used to speak
11. used to bark

Quiz 18, p. 27
1. Joe **didn't** walk to work yesterday. He **took** the bus.
2. Mary **went** to the emergency room at midnight last night.
3. While Dr. Dann **was listening** to his patient, his cell phone rang. He **didn't** answer it.
4. Marco **used to** swim every week, but now he prefers working out at the gym.
5. After Evie **got** a new job, she **celebrated** her success with her friends.
6. When the phone **rang** at 11:00 last night, I was in a deep sleep. I almost **didn't** hear it.

Quiz 19, p. 28
1. A: do you get up
   B: wake up, rises, slept, felt
2. catches, caught, was chasing, knocked
3. left, bought
4. A: are you doing
   B: am cleaning, got, were sleeping, made
5. traveled, were traveling, visited, didn't stay, enjoyed
6. watch, watched
7. stepped, was waiting, was
8. live, live / are living, moved, missed, am, like
9. A: Did you read
   B: didn't
   A: went

TEST 1, p. 29
Part A
1. took
2. saw
3. wanted
4. left
5. found
6. hiked
7. were hiking
8. heard
9. stood
10. waited
11. were waiting
12. came
13. was eating
14. saw
15. walked
16. didn't follow
17. got

Part B
A: learned
B: wasn't, taught, showed
A: Do you like
B: love

Part C
1. Did, made
2. did, didn't hear
3. look, am, didn't sleep

Part D
1. Bees used to **make** honey in a tree next to our house until lightning hit the tree.
2. Carol **didn't** go to work yesterday because her son **was** sick.
3. Doug and Peter **had** a party last weekend. Everyone from our class **came**.
4. **Were** you upset with your test results yesterday?
5. After Bill **woke** up, he **got** up.
6. Liz and Ron **planned** to get married last summer, but just before the wedding, Ron **lost** his job. Now they **are waiting** until next summer.

TEST 2, p. 31
Part A
1. was
2. sat
3. didn't go
4. was sitting
5. watched / was watching
6. was talking
7. was trying
8. was crying
9. was eating
10. looked
11. came
12. began
13. waited
14. felt
15. was
16. got

Part B
A: are you doing
B: am looking, went
A: do they live
A: did you do
B: rented, spent, took, were walking, saw

Part C
A: did
B: had
A: Was
B: was eating, came

Part D
1. Dr. Martin used to **work** in a hospital, but now she has a private practice.
2. I was home alone last night. First, I cooked dinner. Then I **washed** the dishes.
3. Matt **was** busy yesterday. He **didn't** go to the party.
4. The Millers **bought** a restaurant last month. They **opened** for business last week.
5. The baby **tried** to crawl a few times, but her legs **aren't / weren't** strong enough yet.
6. Professor Scott **didn't have** time to help us with our lab experiment yesterday. Maybe she can today.

CHAPTER 3

Quiz 1, p. 33
1. tomorrow
2. tomorrow
3. every day
4. yesterday
5. every day
6. yesterday
7. tomorrow
8. every day
9. tomorrow
10. yesterday

Quiz 2, p. 33
1. will have, are going to have
2. will meet, is going to meet
3. will return, am going to return
4. will be, are going to be
5. will fly, are going to fly

Quiz 3, p. 34
1. Will the cat catch
   Is the cat going to catch
2. Will Dr. Brown retire
   Is Dr. Brown going to retire
3. Will your family be
   Is your family going to be

Answer Key 255

4. Will our team win
   Is our team going to win
5. Will Mr. and Mrs. Bell find
   Are Mr. and Mrs. Bell going to find

Quiz 4, p. 35
1. A: are we going to leave
   B: are going to leave
2. A: Are you going to stop
   B: am going to buy
3. A: is not / isn't going to visit
   B: is she going to do
   A: is going to spend
4. A: are going to go
   B: is going to perform
   A: am not going to be

Quiz 5, p. 36
1. will drive
2. will take
3. will not / won't be
4. will listen
5. will be
6. will rain
7. will bring
8. will take
9. Will we go
10. will not / won't have

Quiz 6, p. 36
1. We're
2. we'll
3. I'm
4. You're
5. aren't
6. He's
7. He'll
8. You'll
9. won't
10. We're

Quiz 7, p. 37
1. prediction
2. prior plan
3. decide / volunteer
4. prediction
5. prediction
6. prior plan
7. decide / volunteer
8. prior plan
9. prediction
10. decide / volunteer

Quiz 8, p. 38
1. will
2. is going to
3. are going to
4. will
5. will
6. are going to
7. will / are going to, will
8. is going to, are going to

Quiz 9, p. 39
2. 50%
3. 90%
4. 50%
5. 100%
6. 100%
7. 100%
8. 90%
9. 50%
10. 50%
11. 90%

Quiz 10, p. 39
1. Tom may take the car to the mechanic's tomorrow. OR The mechanic will probably fix / is probably going to fix the problem.
2. Maybe Max will quit / is going to quit his job. OR Max's boss probably won't be / isn't going to be happy about that.
3. Yujung may get a job in international business. OR Maybe she will earn / is going to earn a good salary with her language skills.
4. The Shepherds may buy some paint next weekend. OR They probably won't paint / aren't going to paint their house a dark color.
5. My children probably won't go / aren't going to go to a movie theater this weekend. OR Maybe they will watch / are going to watch a movie at home.

Quiz 11, p. 40
1. Before the Smiths fly to Thailand, they will pick up their airplane tickets.
2. As soon as Sonya gets dressed, she will go to work.
3. Chris will stay home until he feels better.
4. After Mr. Hill takes the driving test, he will get a driver's license.
5. Beatriz will check her answers before she turns in her test paper.
6. When Thomas gets a new phone, he will text me.
7. If Janice wins a lot of money, she will quit her job.
8. David and Sam will go home after they go to the staff meeting.
9. Before Ellen makes lunch, she will wash her hands.
10. Josh will be in bed by midnight if he finishes his homework.

Quiz 12, p. 41
A.
1. am changing
2. are having
3. is joining
4. is taking
5. Are, moving

B.
1. tonight, right now, all next week
2. today, tomorrow, tonight, next semester
3. today, in one hour, next week, next year
4. now, in ten minutes, tomorrow, next week
5. tomorrow, soon, in a few minutes

Quiz 13, p. 42
A.
1. is going, is going to go
2. is visiting, is going to visit
3. leaves, is leaving, is going to leave
4. is staying, is going to stay

B.
1. starts, is starting, is going to start
2. am bringing, am going to bring

C.
1. is having, is going to have
2. opens, is opening, is going to open
3. are going, are going to go
4. closes, is closing, is going to close

Quiz 14, p. 43
2. b
3. a
4. a
5. b
6. a
7. b
8. a
9. a
10. b
11. b

Quiz 15, p. 43

2. am
3. am spending
4. get
5. will find
6. cook
7. am going to look
8. may be
9. will
10. wake
11. probably

Quiz 16, p. 44

1. After I **feed** the children, I will start dinner for the rest of us.
2. In two years, I **will quit** and **sail** around the world.
3. You **will get / 're going to get** a ticket if you continue to drive so fast.
4. Tina **will wash / is going to wash / is washing** the windows this afternoon.
5. Ms. Reed **may arrive** by 10:00 tomorrow.
6. I **am going / am going to go** to a conference on early childhood learning next week.
7. Shhh. The baby **is** about to go to sleep.
8. When **are** we going to leave?
9. If I **have** time, I will help you.
10. Charlie and Kate **are getting / are going to get** married next summer.

Quiz 17, p. 45

1. clears, will take off / is going to take off, will have / are going to have / have
2. gave, were
3. am taking / am going to take, are meeting / are going to meet, finish, are going / are going to go
4. B: will check
   A: am coming
5. A: Are you going to watch / Will you watch
   B: am working / am going to work
6. B: rings
   A: will the bell ring / is the bell going to ring
7. am picking up / am going to pick up, (am) taking / (am going to) take, are eating / are going to eat, is spending / is going to spend, does

TEST 1, p. 46

Part A
1. will explain
2. will probably see
3. won't forget
4. will go, Will you get

Part B
5. A: are you going to do
   B: am going to visit
   A: Are you going to stay
   B: am probably going to be, I am going to look

Part C
6. will help
7. am going to make
8. am going to ask, is she going to say / will she say, she will say / she is going to say

Part D
1. is going to play / is playing
2. starts / will start / is going to start
3. is traveling / is going to travel
4. are driving / are going to drive
5. is going to win / will win

Part E
1. A: crashed
   B: Did you lose
   A: make, will check
2. A: does your flight arrive / is your flight arriving

Part F
1. After we ~~will~~ get married, we will buy a house.
2. Correct
3. Mari will sing in the choir and ~~going to~~ play the piano at her school concert next week.
4. I **am going / will go / am going to go** downtown tomorrow with my friends.
5. My husband and I **are** will not use our credit card so much next month.
6. If Eric **calls** me, I am going tell him I am not available to work this weekend.
7. Correct
8. Tomorrow when John **gets** home, he will help you plant your vegetable garden.
9. Toshi **may** quit his job soon. / **Maybe Toshi will** quit his job soon.
10. Yoko **will cry / is going to cry** when she hears my news.

TEST 2, p. 48

Part A
1. B: will rain, will be, will probably have, won't rain
   A: will bring

Part B
2. A: are you going to do
   B: am going to take, put, am probably going to give
   A: are going to like

Part C
3. will dry
4. am going to go, am going to swim
5. are going to give
6. will get

Part D
1. are planting / are going to plant
2. are going to pull / are pulling
3. will dig up / are going to dig up, are going to decide / will decide
4. will be, is going to be

Part E
A: were you
B: was, needed
A: Are you working / Are you going to work
B: will call

Part F
1. Correct
2. The business office **will close / is going to close** for one week next month.
3. Dinner is almost ready. The oven timer **is** about **to go** off.
4. Fortunately, our teacher **is not going to** give us a quiz tomorrow.
5. Next Saturday, Boris will stay home and **clean** out his garage.
6. Correct
7. Masako **is buying / is going to buy / will buy** a new truck next week.

8. Our electric bill **may increase** next month. OR **Maybe** our electric bill is going to increase next month.
9. Correct
10. Tomorrow when Pierre **gets** to work, he is going to interview several candidates for the assistant manager position.

CHAPTER 4

Quiz 1, p. 50
2. been
3. smiled
4. gone
5. studied
6. come
7. played
8. taken
9. tied
10. stopped
11. made
12. done
13. found
14. known
15. had
16. gone

Quiz 2, p. 51
1. A: Have Cara and Jenn ever taken
   B: haven't, have never done
2. A: Have you ever gone
   B: have, have been
3. A: Has Adam ever played
   B: hasn't, has never played
4. A: Has your teacher ever given
   B: has, have had
5. A: Has Natalia ever made
   B: hasn't, has never cooked

Quiz 3, p. 52
2. Has Miriam picked up her kids at school yet?
   No, she hasn't picked up her kids at school yet.
3. Has Miriam gone to the gym yet?
   Yes, she has already gone to the gym. OR
   Yes, she has gone to the gym already.
4. Has Miriam already had dinner with the Costas? OR Has Miriam had dinner with the Costas already?
   No, she hasn't had dinner with the Costas yet.
5. Has Miriam met with the electrician yet?
   Yes, she has already met with the electrician. OR
   Yes, she has met with the electrician already.
6. Has Miriam gotten a haircut yet?
   No, she hasn't gotten a haircut yet.

Quiz 4, p. 53
2. has been
3. has worked
4. has met
5. has loved
6. has visited
7. has gone
8. has eaten
9. has drunk
10. has seen
11. has traveled

Quiz 5, p.54
2. has enjoyed
3. have been
4. has become
5. has visited
6. has written
7. has added
8. has had
9. has drawn
10. (has) painted
11. has found

Quiz 6, p. 54
1. for
2. since
3. for
4. since
5. for
6. since
7. since
8. since
9. for
10. for

Quiz 7, p. 55
1. since, for
2. for, for
3. since, since
4. for, since
5. since, for, since

Quiz 8, p. 55
1. was
2. took
3. have been
4. received
5. has had
6. met
7. took
8. drove
9. has ridden
10. came

Quiz 9, p. 56
1. a
2. b
3. b
4. b
5. b
6. a
7. b
8. a
9. b
10. b

Quiz 10, p. 57
1. A: haven't eaten, wasn't
   B: have had
2. Has Oscar ever been, has, went
3. tried, haven't had
4. won, has won

Quiz 11, p. 57
A.
1. has been barking
2. has been beeping
3. has been speaking
4. have been driving
5. have been trying

B.
1. How long have you been standing here?
2. I have been working since 10:00 A.M.
3. It has been snowing for two days.
4. How long have they been studying for the final exam?
5. The taxi driver has been waiting for ten minutes.

Quiz 12, p. 59
2. am reading, have been reading
3. aren't working, haven't been working
4. is teaching, has been teaching
5. are dancing, have been dancing
6. isn't studying, hasn't been studying
7. are practicing, have been practicing
8. are doing, have been doing
9. is talking, has been talking
10. am not eating, have not been eating
11. are laughing, have been laughing

Quiz 13, p. 59
1. is doing, has been studying
2. A: are you doing
   B: am watching, have been watching
3. A: are you going, have been shopping
4. A: have been talking
   B: am talking
5. B: haven't been working

Quiz 14, p. 60
1. has gone
2. have been combing
3. haven't talked
4. have been cooking
5. has never flown
6. have known
7. have been working
8. have heard
9. has been working
10. hasn't paid

Quiz 15, p. 61
1. a. 2
   b. 1
2. a. 1
   b. 2
3. a. 2
4. a. 1
   b. 2
5. a. 2
   b. 1

Quiz 16, p. 62
1. had already left
2. hadn't thought
3. had already eaten
4. had already put up
5. had started
6. had already sold
7. had already paid
8. had already met
9. had already read
10. had left

Quiz 17, p. 63
1. Rita **left** two weeks ago to visit her family in the Philippines.
2. … She **hadn't visited** them for two years.
3. Her sister and brother-in-law **have lived** in Manila for many years, too.
4. … She **has been** there many times already.
5. Rizal Park **has** some beautiful Japanese and Chinese gardens
6. Every time she has gone to Manila, she has **taken** her niece to the Star City Amusement Park.
7. … Rita **hasn't** visited it yet.
8. Since they updated the Manila Baywalk several years ago, it **has** become Rita's favorite place in Manila.
9. … Rita **has been / has gone** there often in the evenings.
10. Rita's parents **have** come to the U.S. several times, but Rita always enjoys going "home" to the Philippines.

Quiz 18, p. 64
1. c
2. b
3. d
4. a
5. c
6. c
7. a
8. d
9. b
10. b
11. d
12. c
13. d
14. d
15. c
16. c
17. c
18. a
19. d
20. b

TEST 1, p. 66
Part A
1. paid
2. swum
3. known
4. waited
5. studied
6. told
7. left
8. flown
9. begun
10. eaten

Part B
1. for
2. since
3. since
4. for
5. since
6. for

Part C
1. moved, have met
2. have had, was
3. have played

Part D
1. A: Have you ever tried
   B: haven't, have made
2. A: Have you finished
   B: have
3. A: has been scratching
4. A: Has Anna driven
   B: hasn't
5. A: Have you been crying

Part E
1. a
2. b
3. a
4. b
5. b

Part F
1. Gary **has been** at work since 5:00 this morning.
2. Steve **has** enjoyed listening to Mozart since he took a music class in high school.
3. Nadia **hasn't** finished her dinner yet.
4. It **has rained** on my birthday every year for the last ten years.
5. I **have already decided** to major in marine biology.

TEST 2, p. 68
Part A
1. visited
2. spoken
3. thought
4. given
5. shopped
6. bought
7. come
8. read
9. taught
10. found

Part B
1. for
2. for
3. since
4. since
5. for
6. since

Part C
1. quit, have traveled
2. have been, visited
3. have enjoyed

Part D
1. A: Have you ever worn
   B: haven't
2. A: Have you been jumping
3. A: has been flying
4. A: Have you ever gotten
   B: have, have gotten
5. A: Has Bill done
   B: has

Part E
1. b
2. a
3. a
4. b
5. a

Part F
1. Sandy **has** been trying to call you.
2. Ted **hasn't** called yet.
3. Andy **has been** on vacation since Saturday.
4. I **have known** about those problems for a few weeks
5. Chris **has started** his Ph.D. thesis several times.

CHAPTER 5

Quiz 1, p. 70
1. Do, do
2. Are, am
3. Is, is
4. Did, did
5. Were, weren't
6. Have, haven't
7. Did, did
8. Will, will
9. Is, is
10. Do, do

Quiz 2, p. 71
1. A: Do you like her fiancé?
   B: I do.
2. A: Did they meet in school?
   B: they didn't.
3. A: Have all the wedding plans been made?
   B: they have.
4. A: Are they inviting many people?
   B: they are.
5. A: Will Mr. and Mrs. Jennings be at the wedding?
   B: they won't
6. A: Will you sing at the wedding?
   B. I will.
7. A: Are you nervous?
   B: I'm not.
8. A: Are they going to move to an apartment?
   B: they are.
9. A: Has your sister bought her dress yet?
   B: she hasn't.
10. A: Did I ask too many questions?
    B: you did.

Quiz 3, p. 72
1. a
2. b
3. c
4. a
5. b
6. a
7. a
8. c
9. c
10. a

Quiz 4, p. 73
Suggested answers:
1. Where are Sven and Erik going?
   When are Sven and Erik flying to Greece?
2. What time will their flight arrive in Athens?
   When will their flight arrive in Athens?
3. Where does Sven want to go?
   What does Sven want to go to Athens for?
4. When was Erik a student in Greece?
   How come Erik was in Greece five years ago?
5. Where will Sven and Erik go on July 15th?
   Why will Sven and Erik return home on July 15th?

Quiz 5, p. 74
1. What
2. Who
3. What
4. Who
5. What
6. Who
7. Who
8. What
9. Who
10. Who

Quiz 6, p. 75
1. Who did Bill see?
2. Who saw the bear?
3. Who did Marcella pay?
4. What did Charles order?
5. Who rides a motorcycle?
6. Who came late?
7. What did Ruth bring home?
8. What broke?
9. Who did Laura call?
10. Who won a contest?

Quiz 7, p. 76
1. What do you do
2. What are you going to do / What are you doing
3. What does Anne usually do
4. What did you do
5. What will you do / What are you going to do
6. What would you like to do
7. What do you want to do
8. What are you doing
9. What did Jason do
10. What do geologists do

Quiz 8, p. 77
1. Which
2. What
3. What
4. which
5. what
6. What
7. which
8. What
9. Which
10. What

Quiz 9, p. 78
2. d
3. i
4. g
5. j
6. a
7. e
8. c
9. h
10. k
11. b

Quiz 10, p. 79
1. often, far, long
2. many, long, far
3. many, long, often, many

Quiz 11, p. 80
1. How long did it take to check your email?
2. How do you spell "elephant"?
3. How often do you go to the movies? OR
   How many times a month do you go to the movies?
4. How did you come / get here?
5. How far away is your school? OR
   How many blocks away is your school?
6. How cold is this water?
7. How do you pronounce "Ms."?
8. How old is Jon?
9. How are you feeling?
10. How well does Mr. Wang speak English?

Quiz 12, p. 81

A.
2. Why do you want to work here?
3. Where do you work now?
4. How long have you worked there?
5. What was your favorite project?
6. Who is your supervisor?

B.
1. Who had the most exciting vacation?
2. How far did you ride?
3. How did it feel?
4. When did you go to the beach?
5. How often do you go to the beach?

Quiz 13, p. 82

1. aren't
2. are
3. don't
4. aren't
5. is
6. do
7. isn't
8. does
9. aren't
10. do

Quiz 14, p. 82

1. haven't
2. have
3. wasn't
4. didn't
5. did
6. aren't
7. didn't
8. have
9. hasn't
10. won't

Quiz 15, p. 83

2. b
3. a
4. b
5. b
6. a
7. a
8. b
9. a
10. a
11. b

TEST 1, p. 84

Part A
1. How
2. How long
3. When / How soon
4. Who
5. How long
6. Which
7. Why
8. How often
9. What
10. How far

Part B
1. How often does she train? / How much does she train? / How long does she train?
2. How far does she run? / How much does she run?
3. How fast does she run?
4. When is she going to run in a ten-kilometer race?
5. Who will she run with / With whom will she run?
6. What does she plan to do?
7. How does she spell her name?

Part C
1. don't
2. did
3. hasn't
4. are
5. didn't
6. isn't

Part D
1. What **do** you know about the new department manager?
2. Why **is that dog barking** so loudly?
3. What **does "besides" mean**?
4. Marta changed jobs last month, **didn't** she?
5. Which movie **did** you see last night, *Monsters* or *Dragons*?
6. **How far is it** from Paris to London? / **How long does it take to go** from Paris to London?
7. **Who told** you about the party?

TEST 2, p. 86

Part A
1. Where
2. How far
3. How long
4. Why
5. How often
6. How
7. Who
8. When / How soon
9. Which
10. When / How soon

Part B
1. When did Jill get a new job?
2. Where does she work?
3. What is she going to do for the company?
4. How long does she plan to be there?
5. How often does she drive to work?
6. How does she go to work on the other days?
7. How many hours a week does she work? OR How much does she work?

Part C
1. didn't
2. is
3. do
4. didn't
5. isn't
6. are
7. do

Part D
1. What **does "anyway" mean**?
2. Sonya needs more time, **doesn't** she?
3. What kind **of** soup **do** you want for lunch?
4. **How do you say** "good morning" in Spanish?
5. **Why were you** late today?
6. How long **does it take** to get from here to your home?

CHAPTER 6

Quiz 1, p. 88

2. mice
3. leaf
4. cities
5. tomato
6. boxes
7. women
8. sheep
9. teeth
10. children
11. businesses

Quiz 2, p. 88

2. /əz/
3. /s/
4. /z/
5. /z/
6. /s/
7. /əz/
8. /s/
9. /z/
10. /s/
11. /əz/

Answer Key 261

Quiz 3, p. 89

1. | Steve | asked | a question |
       S   V   O
2. | His question | wasn't | Ø |
       S   V   O
3. | My phone | rang | Ø |
       S   V   O
4. | I | answered | the phone |
       S   V   O
5. | Hanifa | loves | animals |
       S   V   O
6. | She | has had | many pets |
       S   V   O
7. | The police | stopped | several cars |
       S   V   O
8. | The drivers | looked | Ø |
       S   V   O
9. | My daughter | goes | Ø |
       S   V   O
10. | She | is studying | anthropology |
        S   V   O

Quiz 4, p. 90

1. a. N
   b. V
2. a. V
   b. N
3. a. V
   b. N
4. a. N
   b. V
5. a. V
   b. N

Quiz 5, p. 90

Preposition   *Object of preposition*

1. to   school
   during   the winter months
2. in   the snow
   at   each other
3. with   many people
   in   the mountains
4. into   a frozen lake
   near   their home
   under   the ice
   to   safety

Quiz 6, p. 91

2. at
3. in
4. on
5. in
6. on
7. in
8. in
9. on
10. in
11. on

Quiz 7, p. 91

1. lives
2. are
3. is
4. needs
5. don't
6. use
7. are
8. have
9. Do
10. are

Quiz 8, p. 92

1. funny
2. big, good
3. comfortable
4. hot, buttery
5. refreshing, cold
6. long, late

Quiz 9, p. 92

1. He used a small laptop to check new messages.
2. Nan put the clean clothes into the lower drawer.
3. The funny story was about a friendly monster.
4. The worried manager looked at the angry workers.
5. The happy child played with her favorite toy.

Quiz 10, p. 93

1. Your **flower** garden has many unusual flowers.
2. The mosquitos were really bad on our camping trip. I got a lot of **mosquito** bites.
3. There is **customer** parking in front of the store. The customers are happy about that.
4. I see three **spider** webs in the bathroom. I hate spiders!
5. All the **computer** printers in the library are new, but the computers are old.
6. Three people in our office are celebrating their birthdays tomorrow. There will be a lot of **birthday** cake to eat.
7. Don't throw away the **egg** cartons. We will put the hard-boiled eggs in them.
8. I love all the noodles in this soup. It's great **noodle** soup.
9. My doctor gave me an **exercise** plan. I have to do my exercises every day.
10. Collin lives in a two-**bedroom** apartment. The bedrooms are quite large.

Quiz 11, p. 93

2. them
3. They
4. We
5. We
6. us
7. We
8. them
9. She
10. her
11. her

Quiz 12, p. 94

1. teachers'
2. Carlos's
3. wife's
4. children's
5. hospitals'
6. woman's
7. grandson's
8. students'
9. theater's
10. city's

Quiz 13, p. 95

A.
1. Who's
2. Who's
3. Whose
4. Who's
5. Whose

B.
1. Who's the new neighbor?
2. Whose dog always barks at the mailman?
3. Whose car is parked across the street?
4. Who's washing windows?
5. Who's coming to the barbecue?

Quiz 14, p. 96
1. Mine
2. her
3. mine
4. your, ours
5. its
6. it's
7. Their, They're, theirs, it's, their
8. our, It's, it

Quiz 15, p. 96
1. himself
2. yourselves
3. herself
4. myself
5. herself
6. ourselves
7. himself
8. itself
9. themselves
10. yourself

Quiz 16, p. 97
1. another
2. the other
3. another
4. another
5. the other, the other
6. another, the other
7. another
8. the other

Quiz 17, p. 97
1. the others
2. other
3. the others
4. others
5. other
6. others
7. the other
8. the others
9. the others
10. others

Quiz 18, p. 98
1. the other
2. another
3. The others
4. others
5. another
6. others
7. another
8. the other
9. another, other

Quiz 19, p. 99
1. There are 30 **days** in the month of April.
2. Our **apartment** manager is out of town this week.
3. **Whose** glasses are on your desk?
4. She was born **on** September 8, 2003.
5. Mr. **Lee's** company recycles old computers.
6. The cars in the city **make** the streets very noisy.
7. Many people are waiting in the rain to buy tickets for the concert. Everyone **seems** patient, but cold.
8. The **children's** swimming pool in the city is open to all children.
9. I broke my right hand, so I need to write with **the other** one.
10. The dancers practiced **their dance steps in the studio all morning.**

TEST 1, p. 100

Part A
1. phone's, instructions
2. daughters', names
3. Today's, articles
4. Beth's, computers
5. speaker's, ideas

Part B
1. yours, Mine
2. his, her, herself
3. yours, hers
4. me, They're, their, there, It's
5. me, We

Part C
1. the others
2. another
3. other
4. others
5. Another, The other

Part D
1. a
2. b
3. a
4. b
5. a

TEST 2, p. 102

Part A
1. friend's, puppies
2. Brown's, classes
3. boys, babies'
4. Students, university's, problems

Part B
1. mine
2. Your, yourself
3. its, its, It's
4. His, He, himself, him, His
5. I, his
6. there, They're

Part C
1. another
2. other
3. another
4. the others
5. others
6. The other

Part D
1. b
2. b
3. a
4. b
5. a

CHAPTER 7

Quiz 1, p. 104
1. is able to ride
2. couldn't find
3. may be
4. can help
5. may borrow
6. come
7. have to try
8. will not buy
9. aren't able to eat
10. have to leave

Quiz 2, p. 104

A.
1. can't, can
2. can't, can
3. can, can't
4. can, can't
5. can, can't

B.
1. could / was able to
2. couldn't / wasn't able to
3. could / was able to, couldn't / wasn't able to
4. could / was able to

Quiz 3, p. 105
2. permission
3. possibility
4. permission
5. possibility
6. possibility
7. permission
8. permission
9. possibility
10. permission
11. possibility

Quiz 4, p. 106
1. We might go away this weekend.
2. Maybe it will snow tomorrow.
3. Maybe our basketball team will win the championship.
4. Joan might be in the hospital.
5. David may take the driving test tomorrow.
6. Sara might go shopping with us this afternoon.
7. Maybe James will be late for the meeting.
8. My keys may be in my backpack.
9. We may watch a movie tonight.
10. Maybe the car will be ready later this afternoon.

Quiz 5, p. 107
2. past ability
3. present possibility
4. past ability
5. future possibility
6. present possibility
7. future possibility
8. past ability
9. future possibility
10. present possibility
11. future possibility

Quiz 6, p. 108
1. might
2. can
3. can
4. may
5. may
6. could
7. can't
8. can
9. Maybe
10. couldn't

Quiz 7, p. 109
1. May, Could, Can
2. Could, Can, Would, Will
3. Could, Can, Would, Will
4. May, Could, Can
5. May, Could, Can
6. Could, Can, Would, Will
7. May, Could, Can
8. could, can, would, will
9. May, Could, Can
10. Could, Can, Would, Will

Quiz 8, p. 110
A.
1. should be
2. should learn
3. shouldn't expect
4. shouldn't be
5. should try

B.
Answers will vary.
1. You **should / ought to** take it to the repair shop.
2. You **should / ought to** go to the doctor.
3. You **should / ought to** get something to eat.
4. He **should / ought to** talk to his teacher about it.
5. You **should / ought to** get another cup of hot coffee.

Quiz 9, p. 111
A.
1. should / ought to
2. had better not
3. should / ought to
4. had better
5. shouldn't

B.
Answers will vary.
1. should / ought to / had better put cold water on it
2. should / ought to / had better call for help
3. should / ought to / had better call the apartment manager
4. should / ought to / had better call his friend
5. should / ought to / had better fix it before someone gets hurt

Quiz 10, p. 112
1. had to
2. had to
3. have to
4. must
5. have got to
6. have to
7. have got to
8. must
9. had to
10. has got to

Quiz 11, p. 112
1. don't have to
2. must not
3. don't have to
4. must not
5. don't have to
6. don't have to
7. doesn't have to
8. don't have to
9. must not
10. must not

Quiz 12, p. 113
1. must / has to ask
2. must have
3. must not forget
4. don't have to worry
5. had to go
6. didn't have to take
7. have to buy
8. doesn't have to pay
9. had to be
10. had to travel

Quiz 13, p. 114
2. necessity
3. logical conclusion
4. necessity
5. necessity
6. logical conclusion
7. logical conclusion
8. necessity
9. logical conclusion
10. necessity
11. logical conclusion

Quiz 14, p. 114
1. must
2. must not
3. must
4. must
5. must not
6. must not
7. must
8. must
9. must not
10. must

Quiz 15, p. 115
1. don't
2. shouldn't
3. can
4. does
5. won't
6. should
7. wouldn't
8. won't
9. could
10. would

Quiz 16, p. 116
1. Prepare
2. Measure
3. Mix
4. Add
5. Pour
6. Bake
7. test
8. let
9. take
10. decorate

Quiz 17, p. 117
Check 3, 4, 6, 7, 9, 11.

Quiz 18, p. 117
1. would rather
2. likes
3. prefer
4. would rather
5. prefers
6. would rather
7. prefers
8. like
9. prefer
10. would rather

Quiz 19, p. 118
2. b
3. c
4. b
5. a
6. a
7. c
8. b
9. c
10. a
11. c
12. a
13. a
14. a
15. c
16. a
17. b
18. a
19. b
20. c
21. c

TEST 1, p. 120
Part A
1. b
2. c
3. a
4. a
5. c
6. b
7. c
8. b
9. c
10. b

Part B
1. can
2. could / might
3. could
4. could / might
5. could

Part C
1. I had to go to work.
2. I must pass a test.
3. Children shouldn't play with medicine.
4. I should rest.
5. I have to brush my teeth.

Part D
1. I'm feeling hot. **I ought to take** my temperature.
2. **Could / Can / May I borrow** your pen? Mine isn't working.
3. I don't feel like cooking. **Let's order** a pizza.
4. We'll be free on Saturday. We **could meet** then.
5. Look at the sky. It could snow tomorrow, **couldn't it**?
6. Thomas is late. He **could / might / may have** car trouble again.
7. Children **had better not / mustn't play** with matches. They can start fires.
8. **Why don't we go** for a walk after dinner? It's such a nice evening.
9. I don't want to stay home this weekend. **I would rather go** hiking.
10. Jenny **has to be** more careful with her glasses. She has broken them twice.

TEST 2, p. 123
Part A
1. b
2. a
3. c
4. b
5. b
6. b
7. a
8. c
9. a
10. c

Part B
1. can
2. could
3. might / could
4. might
5. could

Part C
1. I must go to the dentist.
2. I should listen to my teacher.
3. People shouldn't steal.
4. I have to brush my teeth.
5. I had to pay some bills.

Part D
1. My grades are low. **I had better study** more.
2. **Can / Could / Would** you please open the window? OR **Please open** the window.
3. I want to stay home tonight. **Let's invite** some friends over.
4. We **can't come** to your party.
5. **Maybe Susan has** a solution to the problem. OR **Susan may have** a solution to the problem.
6. Jackie isn't here. She **could / might / must** be at home in bed.
7. You **mustn't walk** in mud puddles.
8. Why **don't we** go out for dinner tonight?
9. You have to study tonight, **don't you**?
10. I need to make a call. **Could / Can / May** I borrow your phone for a minute?

CHAPTER 8

Quiz 1, p. 125
1. Beth planned to serve pizza, green salad, and ice cream at the party. OR
   Beth planned to serve pizza, green salad and ice cream at the party.
2. Sam, Jeff, and Ellen helped with the decorations. Bob picked up the pizza and drinks. OR
   Sam, Jeff and Ellen helped with the decorations. Bob picked up the pizza and drinks.
3. The party started at 7:00 P.M. Several guests were late.
4. A few people talked, others played games, and several people danced. OR
   A few people talked, others played games and several people danced.
5. Everyone had a wonderful time. No one wanted to go home.
6. Beth thanked everyone for coming and promised to have another party. Then she told everyone good night.

Quiz 2, p. 125
1. B: … Cathy**, and** I … / … Cathy **and** I …
2. A: … ago**, but** he…
   B: … his work email **or** his home email …
3. B: … bicycling**, and** bird watching … / … bicycling **and** bird watching …
   A: … golf**, but** it…
   B: … soccer **and** swimming …

4. A: … spaghetti **or** a grilled cheese sandwich …
      B: … a grilled cheese sandwich **and** a glass of milk, …
   5. B: … a tent, **so** we … / a tent, **and** we …, … campfire, **but** the …

Quiz 3, p. 126
1. a. but
   b. but
   c. so
2. a. so
   b. but
   c. so
3. a. but
   b. so
4. a. but
   b. but

Quiz 4, p. 126
1. doesn't
2. am
3. haven't
4. do
5. don't
6. have
7. are
8. will
9. didn't
10. won't

Quiz 5, p. 127
2. doesn't
3. does
4. is
5. are
6. does
7. do
8. isn't
9. don't
10. aren't
11. is

Quiz 6, p. 128
2. will too
3. isn't either
4. is too
5. neither is
6. so does
7. doesn't either
8. does too
9. neither does
10. so has
11. is too

Quiz 7, p. 129
1. Because Sue really likes sport shoes, she spends a lot of money on shoes.
   Sue spends a lot of money on shoes because she really likes sports shoes.
2. Because Cindy was driving too fast, she had an accident.
   Cindy had an accident because she was driving too fast.
3. Because it was a beautiful day, we went to the beach.
   We went to the beach because it was a beautiful day.
4. Because my old jeans have holes in them, I need to get new jeans.
   I need to get new jeans because my old jeans have holes in them.
5. Because my car is making strange noises, I feel uncomfortable driving.
   I feel uncomfortable driving because my car is making strange noises.

Quiz 8, p. 129
1. so
2. Because
3. because
4. so
5. because
6. so
7. Because
8. because
9. so
10. Because

Quiz 9, p.130
A.
1. b
2. a
3. a
4. b
5. a

B.
1. a
2. a
3. b
4. a
5. b

Quiz 10, p. 132
1. no change
2. Because the students felt the building shake, they got under their desks.
3. Although there was a lot of traffic, we got home on time.
4. no change
5. no change
6. Because my parents were celebrating their 25th anniversary, they had a big party.
7. no change
8. Even though it was raining, we stood in line to get tickets for the concert.
9. no change
10. Although this computer is new, it starts up slowly.

Quiz 11, p. 132
Possible answers:
1. I enjoy **science. My** favorite subjects are **physics, math, and** chemistry.
2. Julia doesn't participate in **sports, and neither do** her friends. OR
   Julia doesn't participate in **sports, and her friends don't either**.
3. Our baseball team lost the **game because** not enough players showed up.
4. My phone isn't working well, and Jack's **isn't** either. OR
   My phone **doesn't work** well, and Jack's doesn't either.
5. I wore a hat and sunglasses at the beach, **and so did** my sister.
6. My mother is Australian, **and / but** my father is Brazilian.
7. Even though you're upset **now, you'll** understand our decision in a few days. OR
   You're upset **now, but you'll** understand our decision in a few days.
8. **Because** our parents both work, my brothers and I sometimes cook dinner. OR
   Our parents both work, **so** my brothers and I sometimes cook dinner.
9. I have never been to Hawaii, and my husband **hasn't either**. OR
   I have never been to Hawaii, and **neither has my husband**.
10. The photographs turned out wonderfully, but the video **didn't**.

TEST 1, p. 133

Part A
1. Elena decided the weather was too nice to stay at **home, so** she packed a picnic lunch and drove to the **beach.**
2. **Even** though it was **crowded, she** found a place to **sit. She** spread out her blanket and opened her lunch **box.**
3. **Inside** was a **sandwich, potato chips**(,) and an **apple. Because** she was still full from **breakfast, she** ate only a little and saved the rest for **later.**
4. **She** took out a book and opened **it. Minutes** later she was **asleep, and** she woke up just as the sun was going **down.**

Part B
1. so
2. Even though
3. but
4. Because
5. or
6. and
7. Even though
8. or
9. so
10. because

Part C
1. so does
2. is too
3. isn't either
4. neither does
5. so do
6. does too
7. neither has
8. will too
9. so are
10. do too

Part D
Possible answers:
1. I study hard **even though** my classes are very easy.
2. After the accident, my left arm hurt, **and so did** my right shoulder. OR
   After the accident, my left arm hurt, **and my right shoulder did too.**
3. Blackberries, strawberries, **and blueberries all** grow in our garden.
4. Because the parking fees were so **high, people** didn't want to park there.
5. We were excited about the concert, **so** we got there early to get good seats.
6. Kate was hungry, **so** she ate a sandwich.
7. Georgie will drive to Seattle, and **so will** Ken.
8. Although I was tired, **I** stayed up late and finished my homework. OR
   I was tired, but I stayed up late and finished my homework.

TEST 2, p. 135

Part A
1. Ron needs to decide if he is going to go to graduate **school, or** if he is going to get a **job.**
2. **He** will finish business school in a few **months. Although** he has enjoyed being a **student, he** is looking forward to **graduation.**
3. **His** parents want him to get a master's **degree. They** have said they will pay for **it, so** they think he should agree to stay in **school.**
4. Ron appreciates their **generosity, but** he also wants to be more independent at this time in his **life. He** wants to start earning his own **money.**

Part B
1. or
2. Because
3. but
4. so
5. Even though
6. because
7. but
8. Even though
9. or
10. so

Part C
1. so is
2. do too
3. so do
4. is too
5. so are
6. so will
7. don't either
8. so did
9. do too
10. neither do

Part D
Possible answers:
1. People couldn't describe the **accident because** it happened so quickly.
2. **Nadia is a new student, but** she has made many friends. OR
   Even though Nadia is a new student, she has made many friends.
3. **A storm was approaching, so** the sailors decided to go into shore.
4. You can pay either by **cash or check.** Which do you prefer?
5. Maria didn't understand the lecture. **Neither did I.**
6. My mom grew up in a small **town, so** she doesn't like to drive in the city.
7. I won't be home **tonight, and neither will my wife.**
8. Harry was hungry at 9:00 A.M. **because** he didn't eat breakfast before school.

CHAPTER 9

Quiz 1, p. 137
1. larger
2. better
3. more dangerous
4. heavier
5. longer
6. more expensive
7. easier
8. quicker
9. more exciting
10. safer

Quiz 2, p.137
1. Summer is warmer than winter.
2. A tiger is bigger than a cat.
3. A day at the beach is more relaxing than a day at work.
4. A comedy is funnier than a drama.
5. Snow is colder than rain.
6. Strawberries are sweeter than lemons.
7. A car is more expensive than a bicycle.
8. A rock is harder than a flower.
9. Taking tests is more stressful than doing homework.
10. A year is longer than a month.

Quiz 3, p. 138
1. the friendliest
2. the biggest
3. the windiest
4. the best
5. the most expensive
6. the smartest
7. the largest
8. the worst
9. the most important
10. the most comfortable

Quiz 4, p. 139
2. I am, they are
3. ours
4. they can
5. hers
6. he can
7. we did
8. he is
9. they do
10. he does

Quiz 5, p. 140
A.
1. the most beautiful, of
2. the best, in
3. the fastest, ever
4. the hardest, of
5. the busiest, in

B.
1. the nicest people
2. the coldest places
3. the worst days
4. the most beautiful paintings
5. the most comfortable beds

Quiz 6, p. 141
1. a. hungrier than
   b. the biggest
2. a. the most helpful
   b. more difficult than
3. a. the most expensive
   b. cheaper than
4. a. older than
   b. the most beautiful
5. a. the coldest
   b. wetter than

Quiz 7, p. 141
1. faster
2. harder
3. more carefully
4. the earliest
5. better
6. more quickly
7. the hardest
8. earlier
9. the best
10. more slowly

Quiz 8, p. 142
1. farther / further
2. further
3. farther / further
4. further
5. farther / further
6. farther / further
7. further
8. farther / further
9. further
10. farther / further

Quiz 9, p. 142
1. hotter and hotter
2. more and more confused
3. harder and harder
4. redder and redder
5. more and more difficult
6. more and more relaxed
7. happier and happier
8. more and more expensive
9. more and more exciting
10. better and better

Quiz 10, p. 143
A.
1. The thicker, the better
2. The faster, the more difficult
3. The colder, the worse
4. The harder, the higher
5. The nicer, the more expensive

B.
1. The darker it got, the more afraid the children were.
2. The more I exercised, the more energetic I felt.
3. The funnier the story was, the harder the children laughed.
4. The later Bob arrives, the more impatient his boss gets.
5. The faster he works, the happier his boss feels.

Quiz 11, p. 144
1. very
2. much / a lot / far
3. very
4. much / a lot / far
5. much / a lot / far
6. very
7. very
8. much / a lot / far, much / a lot / far
9. much / a lot / far

Quiz 12, p. 145
1. a
2. b
3. b
4. a
5. b
6. a
7. a
8. a
9. b
10. a

Quiz 13, p. 146
A.
2. not as warm as
3. almost as warm as
4. just as warm as
5. not as warm as
6. not as warm as

B.
2. Maria
3. Paulo
4. Susan
5. Susan and Lacey
6. Maria

Quiz 14, p. 147
Possible answers:
1. Light chocolate is just as delicious as dark chocolate.
2. Spring is not as colorful as fall.
3. Sending email is not as easy as text messaging.
4. A hard pillow is not as comfortable as a soft pillow.
5. A non-poisonous snake is not as scary as a poisonous snake.
6. Eating is just as important as sleeping.
7. History is not as difficult as biology.
8. A bicycle is not as fast as a motorbike.
9. Basketball is not as popular as soccer.
10. The Winter Olympics are just as exciting as the Summer Olympics.

Quiz 15, p. 148
1. not as sweet as
2. both
3. both
4. both
5. not as fast as
6. both
7. not as soft as
8. both
9. not as fresh as
10. not as hard as

Quiz 16, p. 148
1. more water
2. faster
3. more time
4. more difficult
5. sunnier
6. more problems
7. easier
8. more rain
9. better
10. more quietly

Quiz 17, p. 149
A.
1. to, Ø
2. to, from
3. from, as, Ø
4. Ø, Ø

B.
2. the same / alike
3. different
4. different from
5. similar to
6. similar

Quiz 18, p. 150
1. different
2. the same as / like
3. similar / the same, the same / alike
4. the same
5. different
6. the same / alike
7. different
8. similar / the same
9. the same / alike

Quiz 19, p. 151
1. **The friendliest** person in our class is Julie.
2. The food at the school cafeteria is **worse** than a lunch from home. OR
The food at the school cafeteria is **not as good as** a lunch from home.
3. The movie was **funnier** than we expected. We laughed the whole time.
4. Anna's dog is **the** ugliest dog I've ever seen.
5. Grandpa's behavior is embarrassing. The older he gets, the **more** loudly he talks.
6. I have **the** same bag as you. Where did you buy yours?
7. Venice is one of the most interesting **cities** I've ever visited.
8. For Jon, the game of chess is **like** an interesting math puzzle.
9. As the horse got tired, he began walking slower and **slower**.
10. The driving test was **harder** than I expected.

TEST 1, p. 152

Part A
1. heavier than
2. smaller
3. colder than, the hottest
4. worse
5. the largest, the most crowded
6. easier ... than
7. the biggest
8. the most delicious

Part B
Possible answers:
1. Circle B is as big as Circle D.
2. Circle A is not as big as Circle C.
3. Circle E and Circle C are different.
4. Circle E is the smallest. OR Circle C is the biggest.
5. Circle E is almost as big as Circle A.

Part C
1. of
2. as
3. in
4. from
5. Ø
6. to
7. as
8. of

Part D
1. a
2. c
3. b
4. b
5. c
6. a
7. c

Part E
1. Let's buy this chair. It's less expensive **than** that one.
2. My brother is shorter than **me/I am**.
3. Linda is in **the** same German class as I am.
4. What was **the happiest** day in your life?
5. The kids yelled **more and more loudly** in the park.
6. If you need **further** help, please ask.
7. My homework isn't as difficult **as** yours.
8. Erin's stomachache got **worse and worse** as the day went on.
9. That was one of the best **books** I have ever read.
10. Those students are the **smartest** kids in the class.

TEST 2, p. 155

Part A
1. spicier than
2. the most lovely
3. longer, shorter than
4. more interesting than
5. the biggest
6. the worst, the most famous
7. the most dangerous
8. friendlier than

Part B
Possible answers:
1. Tree A and Tree B are similar.
2. Tree A is just as short as Tree E.
3. Tree C is not quite as tall as Tree D.
4. Tree C is taller than Tree B.
5. Tree A and Tree E are the same.

Part C
1. as
2. in
3. from
4. Ø
5. of
6. to
7. in
8. of

Part D
1. b
2. a
3. c
4. b
5. c
6. c
7. b

Part E
1. These peas are delicious. Fresh peas taste so much better **than** frozen peas.
2. The flu can be a dangerous illness. It's **more** dangerous to have the flu than a cold.
3. Pluto is the smallest **of** the nine planets.
4. The clouds look dark. Let's hope it's not as rainy today **as** it was yesterday.
5. Who has a **better** life: a married person or a single person?
6. I'm a morning person. I have **the most energy** in the morning.
7. As we skied down the mountain, we went **faster and faster**.
8. The ostrich is **the** largest bird, but the elephant is the largest land **animal**.
9. Titanium is one of the strongest **metals** in the world.
10. Please talk more **quietly** in the library.

CHAPTER 10

Quiz 1, p. 158
1. active
2. passive
3. active
4. passive
5. passive
6. passive
7. passive
8. active
9. passive
10. passive
11. active

Quiz 2, p.158
1. is
2. are
3. was
4. were
5. has been
6. have been
7. will be
8. will be
9. is going to be
10. are going to be

Quiz 3, p. 159
1. is driven
2. was hit
3. will be fixed
4. are cleaned
5. have been checked
6. were written
7. was eaten
8. was enjoyed
9. are going to be surprised
10. was signed

Quiz 4, p. 160
A.
1. b
2. a
3. a
4. b
5. a

B.
1. Children enjoy ice cream.
2. The students discussed the book.
3. The new owners have changed the café's name.
4. Our team is going to win the game.
5. The city will give an award to Mr. Reed. OR The city will give Mr. Reed an award.

Quiz 5, p. 161
1. a. Hospital visitors are greeted by volunteers.
   b. Are hospital visitors greeted by volunteers?
2. a. Many different languages are spoken by patients.
   b. Are many different languages spoken by patients?
3. a. Adam's blood pressure has been checked by the nurse.
   b. Has Adam's blood pressure been checked by the nurse?
4. a. A new medication was discussed by the doctors.
   b. Was a new medication discussed by the doctors?
5. a. Excellent care is given by the hospital.
   b. Is excellent care given by the hospital?

Quiz 6, p. 162
A.
2. The furniture is being
3. I am being
4. The carpets were being
5. Paul was being
6. The windows were being

B.
2. Mr. Lim was buying gifts.
3. My grandmother is baking special cakes.
4. Aunt Lily and Uncle Ted are inviting the whole family over.
5. Aunt Lily is preparing a big dinner.
6. Teresa and Louise are setting the table.

Quiz 7, p. 163
2. transitive
3. transitive
4. intransitive
5. intransitive
6. transitive
7. intransitive
8. intransitive
9. transitive
10. transitive
11. transitive

Quiz 8, p. 163
1. no change
2. Our sink has finally been fixed by the plumber.
3. no change
4. no change
5. The car will be sold by our neighbors next month.
6. The dishes are usually washed by Mr. LeBarre.
7. no change
8. no change
9. The broken lamp was returned to the store by Val.
10. no change

Quiz 9, p. 164
1. I was given this sweater. OR This sweater was given to me.
2. Google was created by Larry Page and Sergey Brin.
3. Books are checked out at a library.
4. Has the report been written yet?
5. The picture was painted by Picasso.
6. These walls will be painted tomorrow.
7. The speeding car was stopped by the police.
8. When were cell phones first used?
9. French and English are spoken in Canada.
10. The basketball game has been stopped by the referee.

Quiz 10, p. 165
1. can be contacted
2. has to be taken
3. shouldn't be eaten
4. ought to be sent
5. has to be changed
6. should be given
7. could be stolen
8. must be told
9. might be won
10. may be reached

Quiz 11, p. 166
1. has to be paid
2. has eaten
3. called
4. should be washed
5. will be held
6. can contact
7. was painted
8. was built
9. turned off
10. must be told

Quiz 12, p. 167
1. with / by
2. with / in
3. with
4. in
5. of
6. about
7. in
8. for
9. to
10. with

Quiz 13, p. 167
1. a. is qualified
   b. am excited
2. a. was worried
   b. is bored
3. a. are tired
   b. was crowded
4. a. is pleased / was pleased
   b. is interested
5. a. is engaged
   b. are married

Quiz 14, p. 168
1. a. surprised
   b. surprising
2. a. embarrassing
   b. embarrassed
3. a. interesting
   b. interested
   c. interesting
4. a. amazing
   b. amazed
   c. amazing

Quiz 15, p. 168
1. frightened, frightening
2. excited
3. scary, scared, relaxed
4. disappointing, surprised
5. fascinating, interested

Quiz 16, p. 169
1. get confused
2. got lost
3. got dirty
4. gets nervous
5. getting dark
6. get rich
7. got sunburned
8. got hungry
9. get serious
10. get cold

Quiz 17, p. 170
A.
1. is used to
2. wasn't used to
3. am not used to
4. were used to
5. aren't used to

B.
1. are accustomed to
2. was accustomed to
3. isn't accustomed to
4. is accustomed to
5. am accustomed to

Quiz 18, p. 170
1. used to travel, used to spend
2. are used to eating
3. used to live, Are you used to living
4. is used to being, is used to
5. used to work
6. used to have, am not used to

Quiz 19, p. 171
A.
1. Students are supposed to wear uniforms.
2. Customers are supposed to pay their bills on time.
3. We were supposed to be on time, but we weren't. We were late.
4. I was supposed to make an appointment, but I forgot.
5. It was supposed to snow last night.

B.
Add checkmarks before 2, 4, 5, and 6.
2. Drivers are **supposed** to drive more slowly in rainy weather.
4. You **are** not supposed to wear shoes in the house.
5. We are supposed **to** help Graciela with the dishes.
6. You **weren't** supposed to go to the mall after school.

Quiz 20, p. 172
1. c
2. d
3. a
4. b
5. a
6. b
7. b
8. c
9. d
10. b

TEST 1, p. 174
Part A
1. had
2. spoke
3. were judged
4. were given
5. was
6. were asked
7. wanted
8. argued
9. said
10. was needed / is needed / will be needed

Part B
1. b
2. d
3. d
4. a
5. c
6. b
7. a
8. b

Part C
1. thrilling, frightened
2. disappointed, surprising, confusing
3. amazing, excited

Part D
1. Ben and Rachel **got** engaged last month.
2. Many people are opposed **to** higher taxes.
3. I heard my name. Who **called** me?
4. Dogs in the park **are supposed to be** with their owners.
5. A package **came** a few minutes ago.
6. Our apartment **must be cleaned** before the party next week.
7. I used to **run**, but now I walk for exercise.
8. The fish isn't ready yet. It **should be cooked** a little longer.
9. We enjoyed our time in Malaysia, but **we were exhausted** from the heat.
10. Jorge can skateboard for hours before he gets **tired**.

TEST 2, p. 177
Part A
1. returned
2. had decided
3. was cleaned / had been cleaned
4. were washed / had been washed
5. were dusted / had been dusted
6. thanked
7. looked
8. said
9. was done
10. got

Part B
1. d
2. b
3. a
4. a
5. c
6. d
7. c
8. b

Part C
1. disappointed, boring, exciting
2. shocking, depressed
3. fascinating, interested

Part D
1. The Jeffersons have been married **to** each other for 50 years.
2. My new boss is very interested **in** my work experience.
3. My husband and I used to **live** on a houseboat. Now we rent an apartment downtown.
4. Dr. Barry **arrived** two hours late and missed the meeting.
5. That wallet **is made of** leather. I really like it.
6. The dog began to cross the highway, but there were so many cars that he got **scared**.
7. Pietro is from southern Italy. He isn't used to **driving** in snow.
8. The children are very excited **about** going to the aquarium.
9. Rita's wedding is today. She **is getting** very nervous.
10. What time is the play **supposed to start**?

CHAPTER 11

Quiz 1, p. 180
2. a
3. an
4. a
5. a
6. an
7. a
8. a
9. an
10. a
11. a

Quiz 2, p. 180
2. an
3. some
4. a
5. some
6. some
7. an
8. a
9. a
10. an
11. some

Quiz 3, p. 181
1. s
2. Ø
3. Ø, Ø
4. Ø
5. Ø
6. Ø, s
7. Ø, s
8. Ø, s
9. Ø
10. Ø, es

Quiz 4, p. 181
2. much, Ø
3. much, Ø
4. many, s
5. many, s
6. many, s
7. much, Ø
8. many, s
9. many, s
10. much, Ø
11. much, Ø

Quiz 5, p. 182
1. a little traffic
2. a few apples
3. a little meat
4. a little milk
5. a few coins
6. a little pepper
7. a little vocabulary
8. a few suggestions
9. a few cookies
10. a little money

Quiz 6, p. 183
1. a little
2. much, a little
3. some, a lot of
4. some
5. a little, a lot of
6. much, a lot of, a little
7. many, several, some
8. several, a few
9. some, much, a little
10. much, some

Quiz 7, p. 183
1. coffee
2. chicken
3. irons
4. time
5. light
6. hair
7. glasses
8. some paper
9. works
10. papers

Quiz 8, p. 184
2. bag / box
3. bottle / box / can
4. jar / can
5. box / can
6. bag / box
7. bowl / cup
8. piece / slice
9. cup / glass
10. piece / slice
11. cup / glass

Quiz 9, p. 184
2. non-specific
3. specific
4. specific
5. non-specific
6. non-specific
7. non-specific
8. specific
9. specific
10. non-specific
11. non-specific

Quiz 10, p. 185
1. a. The sunglasses
   b. sunglasses
2. a. bread
   b. The bread
3. a. The furniture
   b. furniture
4. a. children
   b. The children
5. a. Vocabulary
   b. The vocabulary

Quiz 11, p. 185
1. a
2. The
3. the
4. an
5. the
6. a
7. The
8. The, a
9. the

Quiz 12, p. 186
1. the
2. Ø
3. the
4. Ø
5. the
6. the, the
7. Ø
8. The
9. Ø

Quiz 13, p. 187
2. a
3. a
4. a
5. The
6. the
7. the
8. the
9. the
10. the
11. the

Quiz 14, p. 187
1. Ø
2. the
3. Ø
4. The
5. The, the
6. Ø
7. Ø
8. The
9. The

Quiz 15, p. 188
1. an, Ø, a
2. the, the, Ø
3. a, The, a
4. Ø, a / the, the
5. a, Ø, Ø

Quiz 16, p. 189
1. the
2. Ø
3. Ø
4. Ø, the, Ø
5. The, Ø
6. the
7. Ø, Ø
8. Ø, Ø, the, Ø

Quiz 17, p. 189
1. Theresa can't decide whether to study **J**apanese or **C**hinese.
2. **W**here are you going for the summer break?
3. **T**he **A**lps are in **S**witzerland, **A**ustria, and **F**rance.
4. **W**e're reading **S**hakespeare's *Romeo and Juliet* for our literature class.
5. **T**he directions say to turn on **F**ifth **S**treet, but this is **P**ark **A**venue.
6. **L**ast **M**onday was my first day as a student at **S**tanford **U**niversity.
7. **T**he **M**ississippi **R**iver flows into the **G**ulf of **M**exico.
8. **I** was supposed to be born in **A**pril, but **I** was born a month early, so my birthday is in **M**arch.
9. **W**hich instructor do you prefer: **D**r. **C**osta or **P**rofessor **P**eterson?
10. **M**ath 441 is a very high-level math class.

Quiz 18, p. 190
2. b
3. a
4. a
5. d
6. a
7. b
8. c
9. b
10. d
11. a

TEST 1, p. 191

Part A
1. a
2. c
3. c
4. a
5. b
6. b
7. a
8. d
9. b
10. c

Part B
1. The lake is too cold for swimming. **H**ow about going to the indoor pool at **M**ountain **V**iew **P**ark?
2. Our biology class will be taught by **D**r. **J**ones. He's a professor, not a medical doctor.
3. **M**aria's parents are from **M**exico. **S**he speaks **S**panish fluently.
4. **T**he college directors plan to tear down **B**rown **H**all and build a new library.
5. **W**ould you be interested in going on a boat trip down the **C**olorado **R**iver? **W**e would see part of the **G**rand **C**anyon.

Part C
1. a
2. Ø, the
3. a, an
4. The, a
5. a, a, The, the
6. the
7. Ø
8. Ø, the
9. Ø, an, Ø
10. the, the

Part D
1. Let's get a drink of water. **I'm thirsty**.
2. **The weather** at the beach is beautiful in August.
3. Your hair looks great. Did you **get a haircut**?
4. For breakfast, Thomas ordered two eggs and **two slices/pieces of toast**.
5. Here's a map of **the United States**. Do you see California?
6. I need **a little more time** to finish my test.
7. There are no **fish** in the Dead Sea.
8. Aunt Betsy and Uncle Wes are moving **to London** next month.
9. I need to have **my car** checked soon.
10. **Many students** at Shorewood High School study Japanese.

TEST 2, p. 193

Part A
1. b
2. d
3. c
4. c
5. a
6. a
7. d
8. a
9. b
10. c

Part B
1. The assignment for our literature class is to read the first chapter of **S**hakespeare's *Hamlet*.
2. **I** heard that my neighbors plan to visit **E**ngland in **M**ay.
3. **T**omorrow there will be a concert at **W**ashington **P**ark, near **B**roadway **A**venue. **A** music group from **S**outh **A**frica will be playing.
4. **T**here is a **M**iami **U**niversity in **O**hio, but **M**iami is in **F**lorida. **I**sn't that strange?
5. **W**hen did **W**illiam begin working for the **S**onglin **C**orporation?

Part C
1. a, Ø, Ø
2. an
3. the, a, a
4. The
5. Ø, Ø
6. Ø, Ø
7. Ø, Ø
8. a, the
9. the, the
10. The, Ø

Part D
1. I don't **need help** now, but I will later on.
2. There are **a few people** at work who don't like the new manager.
3. **Water** is necessary for life.
4. I married **my brother's** best friend from college.
5. Antoine reached the top of **Mount McKinley** in Alaska yesterday.
6. My friends bicycled through **the Sahara Desert** last summer.
7. Nick worked on his car **for an hour**, but he couldn't fix it.
8. Barcelona is **in Spain**, but it's not the capital city.
9. Carlos texted his dad **several times**, but he didn't get a response.
10. Adrianna has **so much homework** that she stays up late every night.

CHAPTER 12

Quiz 1, p. 195

A.
1. Many tourists **who / that** visit New York City go to Central Park.
2. The French man **who / that** was in my English class loved coffee and chocolate.
3. Tobias thanked the nurse **who / that** took care of him in the hospital.
4. I feel happy around people **who / that** are positive about life.
5. When Maja was on the bus, she sat next to a woman **who / that** was talking on her cell phone.

B.
1. I heard about a teenaged boy who / that takes gifts to children in hospitals
2. Tomas met a marine biologist who / that once swam with sharks.
3. The people who / that work on this boat have fire drills once a month.
4. The police helped an old man who / that was confused and lost.
5. The scientists who / that discovered DNA are very famous.

Quiz 2, p. 196

1. who
2. whom
3. who
4. whom
5. who
6. whom
7. whom
8. who
9. who
10. whom

Quiz 3, p. 196

2. a, b
3. a, b, c, d
4. a, b
5. a, b, c, d
6. a, b
7. a, b
8. a, b, c, d
9. a, b, c, d
10. a, b, c, d
11. a, b

Quiz 4, p. 197

A.
2. a
3. a, b
4. a, b
5. a
6. a, b

B.
1. William got a new winter coat laptop which / that keeps him really warm.
2. The puppy which / that has a lot of energy is black and white.
3. The train which / that arrives exactly at 6:53 P.M. is never late.
4. The ancient city which / that was discovered in 1748 is popular with tourists.
5. Every day, I wear gold earrings which / that were a gift from my husband.

Quiz 5, p. 198

1. I spoke with an amazing woman who is smart, strong and beautiful.
2. Here is the new book which you asked me to order.
3. The young man who crashed into a tree has a broken leg.
4. The cell phone which I bought yesterday has all the newest features.
5. The kind man who owns the gas station repaired my car for free.
6. I don't know the student who wrote an article for the newspaper.
7. The mathematics professor whom / who I met last year is going to retire next month.
8. The documentary which we watched last night was fascinating.
9. The elderly woman who lives in the apartment next to mine has no family.
10. The oranges which are in the bowl on the table are really juicy.

Quiz 6, p. 199

1. get
2. play
3. writes
4. work
5. like
6. is
7. sells
8. are
9. spends
10. serve

Quiz 7, p. 199

1. The music <u>that we listened to</u> was recorded in front of a live audience.
2. The professor <u>that I spoke to / with</u> is not available to teach the class next semester.
3. The company <u>that I work for / at</u> treats its employees well.
4. The people <u>who I depend on</u> the most are my parents.
5. Don't tell me about your plumbing problems. The person <u>whom you should complain to</u> is the building manager.
6. The movie theater <u>we went to</u> had a huge screen.
7. The young man <u>whom Clara was waiting for</u> is a professor at Oxford University.
8. The university <u>that Josh graduated from</u> is in Seattle.
9. The elderly man <u>whom Marta talked to / with</u> told stories about his life.
10. The person <u>whom Greg lives with</u> has both M.D. and Ph.D. degrees.

Quiz 8, p. 200

1. a. which we listen
   b. we listen to
2. a. whom Sebastian works
   b. Sebastian works for
3. a. which I told you
   b. I told you about
4. a. I was looking for
   b. which I was looking
5. a. whom I spoke after class
   b. I spoke to after class

Quiz 9, p. 201

1. The little girl whose doll was stolen was sad for days.
2. I'm friends with a woman whose daughter is training to be a professional boxer.
3. In Kansas City, I met a man whose parents know my grandparents.
4. I have a friend whose sailboat is also her home.
5. I enjoyed meeting the couple whose children go to the same school as our children.
6. The people whose car was just hit are upset.

7. The couple **whose** vacation home we rent has a new baby.
8. I know a woman **whose** work involves designing home security systems.
9. A writer **whose** new book is about mountain climbing spoke about his experiences.
10. The dog **whose** owner left him outside a restaurant is being cared for by the staff.

Quiz 10, p. 202
1. The family **that / who** arrived late discovered they had missed the wedding.
2. A neighbor **whose** son works for an airline can fly anywhere quite cheaply.
3. I ran into a man **who / that** was my boss 20 years ago.
4. Those are the students **who / that** volunteer to clean up parks on weekends.
5. The cookies **which I baked are burned.**
6. The woman **who / whom / that / Ø** I see every day on the bus talks the entire time.
7. I work with a doctor **who / that** has clinic hours two evenings a week.
8. Here is the magazine **that / which** has the story about home theater systems.
9. Aiko and Yutaka moved into the apartment which **is** on the top floor.
10. The people **whose** dog bit the delivery man had to pay the doctor bills.

TEST 1, p. 203
Part A
1. The mail was addressed to our neighbor **who / that** lives in the apartment downstairs.
2. I work with a man whose wife trains police dogs.
3. The garden **which / that / Ø** I had forgotten to water looks healthy again.
4. The pianist **who / that** plays in the hotel lobby on weekends teaches my children piano.
5. The manager **whom / who / that / Ø** I work for treats me fairly.
6. A real estate agent whose name is Mike Hammer called.

Part B
1. which / that / Ø
2. who / that
3. which / that / Ø
4. whose
5. whom / who / that / Ø

Part C
1. are
2. studies
3. live
4. is
5. are

Part D
1. about
2. at / in
3. on
4. from
5. for
6. to

Part E
1. a. he paid a lot of money for
   b. which he paid a lot of money
2. a. whom the taxi was waiting
   b. that / whom / who the taxi was waiting
3. a. I spoke to
   b. whom I spoke

Part F
1. The firefighters **who put** out the fire were dirty and exhausted.
2. The sofa **(that / which) we ordered** still hasn't arrived.
3. The man that **I work for** is blind.
4. Here is the receipt **for** which you asked. OR Here is the receipt **which / that you asked for.**
5. The finger **which / that** I broke is healing well.
6. I studied with a professor **whose** books are well known around the world.
7. I met a little girl whose favorite food **is** mushrooms.

TEST 2, p. 205
Part A
1. The earphones **which / that** I bought didn't work.
2. The couple whose horse won the race was surprised.
3. The supervisor **whom / who / that** people like to work for is very kind.
4. We met a boy whose dog can do a lot of tricks.
5. The actor **who / that** starred in several movies last year won an Academy Award.
6. The little girl picked some of the flowers **which / that** grow in my garden.

Part B
1. which / that / Ø
2. whom / who / that / Ø
3. which / that / Ø
4. whose
5. which / that / Ø

Part C
1. designs
2. wants
3. go
4. speaks
5. barks

Part D
1. to
2. to
3. at / to
4. to / about
5. about
6. to

Part E
1. a. I told you about
   b. I told you
2. a. she had been looking for
   b. which she had been looking
3. a. whom you introduced me
   b. you introduced me to

Part F
1. The pillows Sandra had on her bed **were** too hard.
2. The train **that / which** came through the tunnel blew its whistle.
3. The suitcase **which I bought** has lots of pockets.
4. The doctor **who** operated on my father is very skilled.
5. I work with a woman **who / that** grew up in the same neighborhood as me.
6. The taxi driver **who / that** drove us to the hotel charged too much.
7. Here's an article that you might be interested **in**.

CHAPTER 13

Quiz 1, p. 207
2. opening
3. doing
4. traveling
5. driving
6. washing
7. moving
8. working
9. exercising
10. buying
11. running

Quiz 2, p. 208
1. am going to go / will go / am going shopping
2. went sailing
3. go camping
4. went skydiving
5. go fishing
6. goes jogging
7. go hiking
8. are going to go / will go / are going bowling
9. went sightseeing
10. goes dancing

Quiz 3, p. 208
1. working
2. to work
3. working
4. working
5. to work
6. to work
7. working
8. working
9. to work
10. to work

Quiz 4, p. 209
1. a
2. a, b
3. b
4. a
5. b
6. a
7. a, b
8. a, b
9. a
10. a
11. a
12. b
13. b
14. a
15. a and b
16. a
17. b
18. a
19. b
20. a

Quiz 5, p. 210
1. for
2. of
3. in
4. about
5. on
6. to
7. about
8. like
9. at
10. for

Quiz 6, p. 211
1. by
2. by / with
3. with
4. with
5. by
6. with
7. by / with
8. by, by
9. with

Quiz 7, p. 211
2. by washing
3. by looking
4. by saving
5. by doing
6. by sending
7. by speaking
8. by changing
9. by exercising
10. by working
11. by making

Quiz 8, p. 212
A.
1. Having a quiet place to study is necessary.
2. Studying with friends is fun.
3. Learning a lot of new information is difficult.
4. Taking short breaks is a good idea.
5. Getting a good night's sleep is important.

B.
1. It is important to recycle everything you can.
2. It is helpful to use public transportation instead of your car.
3. It is sensible to turn off lights when you leave a room.
4. It is possible for everyone to save water.
5. It is necessary for all of us to take care of our planet.

Quiz 9, p. 213
A.
1. no change
2. Judy is moving to her hometown **in order** to be closer to her elderly parents.
3. Tom got new glasses **in order** to see better.
4. no change
5. Frankie practiced driving almost every day **in order** to pass the driving test.

B.
1. I turned down the TV in order to hear you better.
2. I wore socks to bed in order to keep my feet warm.
3. I took out a loan in order to buy a car.
4. I called the doctor in order to make an appointment.
5. I turned off my phone in order to get some sleep.

Quiz 10, p. 214
1. to
2. for
3. to
4. to
5. for
6. for
7. to
8. for
9. to
10. to

Quiz 11, p. 215
1. a. too hot
   b. cool enough
2. a. too sour
   b. sweet enough
3. a. enough eggs
   b. too tired
4. a. safe enough
   b. too dangerous
5. a. clean enough
   b. too dirty

Quiz 12, p. 216
1. a
2. a
3. b
4. a
5. a
6. a, b
7. b
8. a
9. a, b
10. b
11. b
12. a
13. a
14. b
15. a
16. b
17. a
18. a
19. a, b
20. a

TEST 1, p. 218
Part A
1. to get
2. reading, to relax
3. to call, to get, to become
4. on getting, to arrive, Being
5. to build / building, to snow / snowing, on making / to make, to have, building, skating

Part B
1. for
2. by
3. at
4. to
5. by
6. by
7. with, with
8. in, about

Part C
1. a. too sick
   b. healthy enough
2. a. cool enough
   b. too hot
3. a. sweet enough
   b. too bitter

Part D
1. b
2. b
3. a
4. b
5. a
6. b
7. b
8. a
9. b

TEST 2, p. 220

Part A
1. working, to start
2. to go
3. at kicking, Playing
4. listening
5. cooking, to rest, making, eating
6. about / of visiting, sightseeing, about speaking, to learn, to travel

Part B
1. in
2. by
3. by, by
4. to
5. by
6. by
7. with
8. about, of

Part C
1. a. loud enough
   b. too quiet
2. a. warm enough
   b. too cool
3. a. too bright
   b. dark enough

Part D
1. a
2. b
3. a
4. b
5. b
6. a
7. b
8. a
9. b

CHAPTER 14

Quiz 1, p. 222
1. I don't know <u>what her name is</u>.
2. Nick doesn't know <u>what time the movie starts</u>.
3. My son wanted to know <u>if I needed help with the dishes</u>.
4. I didn't know <u>that it snowed in April last year</u>.
5. My parents want to know <u>who lives in the big house on the corner</u>.
6. No one knows <u>where Nathan works</u>.
7. My teacher doesn't know <u>whose book this is</u>.
8. James knows <u>that I take the bus to school</u>.
9. Ruth wants to know <u>if I will go to Montreal next summer</u>.
10. I don't <u>know where my keys are</u>.

Quiz 2, p. 222
1. why the Earth has a moon
2. where the sun goes at night
3. when animals go to bed
4. how a chameleon changes its color
5. how tall a giraffe is
6. who planted all the trees in the forest
7. whose trees they are
8. what color the ocean is
9. how many fish live in the ocean
10. where the bottom of the ocean is

Quiz 3, p. 223
1. A. does an apartment cost
   B. rent is
2. A. the conference starts
   B. the date is
3. A. does Hoang have
   B. he has
4. A. is that lady crying
   B. she is crying
5. A. it is
   B. is the jacket

Quiz 4, p. 224
1. if Norma / she is
2. if you have
3. if Oscar / he wants
4. if I'm going
5. if I will have
6. if your brother is coming
7. if Sara / she borrowed
8. if this bus / it goes
9. if he has finished
10. if Joanna and Max / they are

Quiz 5, p. 225
1. The police are trying to prove **that** the bank manager stole the money.
2. My parents were upset **that** I lost my phone.
3. We're disappointed **that** you didn't believe us.
4. Did I tell you **that** we are moving next month?
5. I heard **that** there's a new Korean restaurant in town.
6. For centuries, people were convinced **that** the Earth was flat.
7. Is it true **that** your dad is a movie script writer?
8. Can you believe **that** it's already June?
9. I'm sure **that** "scissors" starts with "S-C."
10. Carlos was impressed **that** Juan knew so much about chemistry.

Quiz 6, p. 225
A.
1. so
2. so
3. not
4. so
5. not

B.
1. I think that Sergei is well enough to return to work.
2. I don't believe that James is telling the truth.
3. I hope that I have learned the names of all of the students.
4. I don't think that I want another cup of coffee.
5. I believe that Amy is going to visit us next month.

Quiz 7, p. 226

1. Carmen asked**, "D**o you have money for parking**?"**
2. The doctor said**, "Y**ou must stop smoking immediately**."**
3. **"**There's a mouse in the house!**"** my mother yelled**.**
4. The policeman said**, "M**ay I see your driver's license**?"**
5. Mary asked**, "D**id you get the message I left for you**?"**
6. The Johnsons said**, "W**e have to leave now**. W**e have another party to attend tonight**."**
7. Our teacher asked**, "W**ho knows the answer**? W**ho would like to write it on the board**?"**
8. My mother said**, "I** won't be home until 7:00 tonight**."**
9. **"**Do you speak Russian?**"** asked Natasha**.**
10. Miguel asked**, "A**re you tired from the walk around Green Lake**?"**

Quiz 8, p. 227

2. "Why not?" I asked.
3. "They're quite ugly," she replied.
4. "Well, they might look unpleasant**,**" I said. "They're not as beautiful as butterflies, but they're good to have around. They eat ants and flies that you don't want to have in your house**,**" I said. "Try to think of them as a gift from nature."
5. "Wow!" she said. "I didn't know spiders were so helpful."
6. "You learn something new every day," I said.
7. "I guess you're right, Dad," she answered.

Quiz 9, p. 227

1. her
2. I, her
3. our, us, them
4. she, my / our, me, my
5. he, his, themselves
6. their, their, They
7. they, their, they, their

Quiz 10, p. 228

1. would be
2. had been
3. had
4. didn't understand
5. hadn't learned
6. was going to start, needed
7. had been
8. could feed, were

Quiz 11, p. 229

1. Abdul said (that) he would be 25 on his next birthday.
2. My parents said (that) they had enjoyed their trip to Costa Rica.
3. My friends said (that) they wanted to have a birthday party for me.
4. Suzanne said that she had lived in Italy for 20 years.
5. The boy said that the dog had taken his ball and (he) wasn't coming back.
6. Emily said that she was going to retire in a few years and that she and her husband were planning to travel.
7. My husband said that he could pick up the kids after school and that I didn't need to worry about it.

Quiz 12, p. 230

1. asked, told
2. said, told
3. told, said
4. said, told
5. said, told, asked, said, told, told, said

Quiz 13, p. 230

1. I don't know **if** the doctor can see you tomorrow or not.
2. Please tell me what **they did**.
3. Do you know whose coat **is on the chair**?
4. Do you know **whether** or not the bus has come? OR Do you know **if** the bus has come **(or not)**?
5. We'd like to know if the subway **stops** here.
6. I **hope that** I can come with you tonight.
7. Leila **said** that she wasn't home last night.
8. I'm sorry **(that)** we have to cancel our plans.
9. The dentist said**, "**Your teeth look very healthy. You are taking good care of them.**"**
10. The teacher isn't sure whether Liz wants help or **not**.

TEST 1, p. 231

Part A

1. where the milk / it is
2. what time we will be finished
3. if / whether there are any eggs left
4. whose homework that is
5. if / whether someone is knocking at the door
6. who called
7. if / whether Kwon finished his biology lab work / it
8. what the weather / it is supposed to be like tomorrow
9. if / whether anyone has met the new coach / him / her yet
10. if / whether we have to type this assignment / it

Part B

1. Mom came in to wake me up. "What time is it?" I asked her.
2. "It's time to get up," she replied.
3. "But it's not a school day," I said. "Please let me sleep in," I begged.
4. "You can't sleep in today," she said. "It's a special day."
5. "What special day**?**" I asked.
6. "It's your birthday," she said.
7. "Oh my gosh! I forgot. I have to get up right away," I said. "I have so many things I want to do today."

Part C

1. Julia said (that) the cookies were ready.
2. The librarian said (that) the library was going to close early today.
3. John asked how far away the airport was.
4. The fire chief said (that) it had taken a long time to put the fire out.
5. The clerk asked me if / whether I wanted to pay with cash or a card.
6. Marika said (that) the flight would arrive in ten minutes.
7. The teacher said (that) the test was going to be on Friday.
8. The students replied (that) they didn't want a test.

9. The manager said (that) we needed to put our cell phones on silent in the office.
10. Joan asked if / whether I had ever watched a movie in Swedish.

Part D
1. My friends understand **what I** like.
2. I'd like to know **if / whether** this computer **works**.
3. The teacher **said / told us** that he would be at a meeting tomorrow.
4. I want to know why **they came**.
5. **It i**s a fact that exercise makes us healthier.
6. Do you know if Rick **lives** here**?**
7. **I'm sure that** we will have a good time together.
8. A classmate asked me **where I lived**. OR
   A classmate asked me, "Where **do you live?**"
9. Do you know whose **keys these are**?
10. I'm not **sure if** he wants to come or not.

TEST 2, p. 233

Part A
1. what the date (today) / it is
2. what year Aaron / he was born
3. if / whether Dimitri and Irina / they got engaged last weekend
4. if / whether anyone has called a plumber yet
5. how many people knew about the problem / it
6. if / whether the weather here / it changes much at different times of year
7. if / whether Paula / she left the company
8. who the new manager / it will be
9. whose car we are taking to the mall
10. if / whether the copy machine / it works

Part B
1. "What do you want to do after you finish school?" my teacher asked.
2. "I'm not sure," I said. "I'd like to have a job that is interesting and pays well."
3. "Everyone would like that," said my teacher. "Is there a specific area you see yourself working in?"
4. "Yes," I replied. "I love working with animals. Maybe I could be a veterinarian."
5. "One way to find out is to work with animals first," said my teacher. "Volunteer at an animal shelter or zoo. See how you like it."
6. I told him, "I like that suggestion. Thanks!"

Part C
1. The reporter asked me if / whether I had time to answer a few questions.
2. Yolanda said (that) she had cleaned her apartment and had done her laundry.
3. My friend said (that) the bus would be late.
4. Joe asked who had taken his car.
5. The manager said (that) they had decided to move their offices to a new location.
6. Brad asked when the book would be published.
7. Shirley said (that) she could fix that for me / us.
8. My parents said (that) they were happy to hear about my new job.
9. The doctor asked me if / whether I had been taking my medicine.
10. The dancers said (that) they were ready for their show.

Part D
1. Can you tell me whose coat **this is**?
2. The doctor said, "**Your** daughter just has a bad cold. It's nothing serious."
3. My friends asked me**,** "**W**hen will you get married?"
   OR
   My friends asked me when **I would** get married.
4. Did my mom ask you **if** you could come to our party **or not**?
5. Hamid asked why **I always came** late**.**
6. Mr. Hill **told me** that he was feeling ill.
7. Could you tell me where Fred **work**s in the evenings?
8. **I think that** you will enjoy being on the soccer team.
9. **I know that** this will be a good opportunity for us.
10. Professor Thomas told us he **would** be absent yesterday.

MIDTERM EXAM 1, p. 235

1. d	11. a	21. c	31. b	41. c
2. a	12. d	22. a	32. d	42. b
3. c	13. d	23. b	33. a	43. c
4. a	14. c	24. c	34. c	44. c
5. d	15. b	25. c	35. b	45. a
6. b	16. c	26. a	36. a	46. b
7. c	17. c	27. c	37. d	47. d
8. a	18. b	28. c	38. c	48. a
9. c	19. b	29. c	39. b	49. c
10. b	20. c	30. b	40. b	50. d

MIDTERM EXAM 2, p. 241

Part A
1. bought
2. gets
3. fell, was
4. is going to start / will start / is starting
5. have never been
6. had taken
7. are drawing
8. doesn't have, takes
9. are going to move / will move / are moving
10. sold
11. has worked / has been working, is going to retire / will retire / is retiring
12. threw
13. had already gotten
14. am going to send / will send, love / will love / are going to love
15. cheered / were cheering, smiled / were smiling

Part B
1. can / may / should / had better / must
2. Can / Could / May
3. could / may / might
4. should / had better / must
5. must
6. could
7. Can / Could / Would
8. should
9. can
10. should / had better / must

Part C
1. Is, what, Do, I do
2. Whose, Do, I do, How, Why, Is
3. Did / Didn't, I didn't, When / What time, Why, How much

Part D
1. **Tom's** last name is Miller.
2. **Tomatoes** are good for us. They have lots of vitamin C.
3. They would rather eat Chinese food, **wouldn't** they?
4. All of the **actors'** names are listed on page six of your program.
5. Do you want to go shopping on Saturday? I need **some** new shoes.
6. Mr. Tobias works in his **flower** garden every day.
7. My brothers started **their** own gardening business last year.
8. Some children prefer to play by **themselves** rather than with other kids.
9. There are two new students in class. One is from Libya, and **the other** is from Romania.
10. Jason forgot **his** math book on the bus. I hope he can get it tomorrow.

FINAL EXAM 1, p. 244

1. b	11. a	21. a	31. c	41. b
2. c	12. c	22. b	32. c	42. c
3. d	13. b	23. a	33. b	43. c
4. a	14. b	24. d	34. c	44. a
5. c	15. d	25. c	35. c	45. b
6. c	16. b	26. d	36. a	46. d
7. b	17. d	27. b	37. a	47. c
8. d	18. c	28. d	38. b	48. d
9. c	19. b	29. b	39. d	49. b
10. d	20. a	30. c	40. c	50. d

FINAL EXAM 2, p. 249

Part A
Possible answers:
1. Mark wants to go to Alaska next summer, but his wife would rather go to California.
2. When Mr. Meecham went to a meeting with a new client, he wore his best suit. OR
Mr. Meecham wore his best suit when he went to a meeting with a new client.
3. This radio station plays lots of classic rock music, so I enjoy listening to it.
4. Chris is a great guitar player, and Monica is a fantastic pianist. OR
Monica is a fantastic pianist, and Chris is a great guitar player.
5. Although Nan and Rose both enjoy playing tennis, they don't play together very often. OR
Nan and Rose both enjoy playing tennis although they don't play together very often.
6. As soon as we finished dinner, we had strawberry shortcake for dessert. OR
We had strawberry shortcake for dessert as soon as we finished dinner.

Part B
2. which
3. who
4. how long
5. that
6. whether (or not)
7. what time

Part C
1. **"I'm** so sorry to hear that your father is ill," said Margaret.
2. John asked, **"D**o you want to have toast or cereal for breakfast?"
3. **"**Could **I** please have a glass of water?" Mary asked. "I'm really thirsty."

Part D
1. Derek said (that) social networking websites were a great way to keep in touch.
2. Heidi said (that) she had always dreamed of being an actress.
3. John asked which video game I liked the best.
4. Anita said (that) she would go to college next year.
5. My friends asked if I wanted to go to the coffee shop with them.

Part E
1. Adam is taller **than** his brothers.
2. **I'll / I can** help you in just a few minutes.
3. Playing the clarinet is not as **difficult** as playing the oboe.
4. That is the **funniest** joke I've heard in a long time.
5. My sisters like country-western music much better than I **do**.
6. If you don't fix the leak in the roof, it will cause **further** water damage.
7. My brother and I are twins, but I **haven't seen** him since 2010.
8. Laptop computers are usually **more convenient** than desktop computers.
9. How many apples **did** you **buy**?
10. My classmate has **the same** name **as** me. We are both named William.

Part F
1. a
2. The
3. An
4. Ø
5. The

Part G
1. How far
2. homework
3. Ø
4. many
5. a little
6. Ø
7. are published
8. the
9. learning
10. Ø
11. several
12. to go
13. have been studying
14. mine
15. was given